Lecture Notes on Data Engineering and Communications Technologies

Volume 143

The aim of the book series is to present cutting edge engineering approaches to data technologies and communications. It will publish latest advances on the engineering task of building and deploying distributed, scalable and reliable data infrastructures and communication systems.

The series will have a prominent applied focus on data technologies and communications with aim to promote the bridging from fundamental research on data science and networking to data engineering and communications that lead to industry products, business knowledge and standardisation.

Indexed by SCOPUS, INSPEC, EI Compendex.

All books published in the series are submitted for consideration in Web of Science.

More information about this series at https://link.springer.com/bookseries/15362

Salah Bourennane · Petr Kubicek
Editors

Geoinformatics and Data Analysis

Selected Proceedings of ICGDA 2022

Editors
Salah Bourennane
Multidimensional Signal Processing Group
Ecole Centrale Marseille
Marseille Cedex 13, France

Petr Kubicek
Department of Geography,
Faculty of Science
Masaryk University
Brno, Czech Republic

ISSN 2367-4512 ISSN 2367-4520 (electronic)
Lecture Notes on Data Engineering and Communications Technologies
ISBN 978-3-031-08016-6 ISBN 978-3-031-08017-3 (eBook)
https://doi.org/10.1007/978-3-031-08017-3

This Springer imprint is published by the registered company Springer Nature Switzerland AG
The registered company address is: Gewerbestrasse 11, 6330 Cham, Switzerland

Preface

Rapid developments in Geoinformatics and Data Analysis and its appalling potential have drawn significant attention in recent years. ICGDA establishes a platform for researchers in Geoinformatics and Data Analysis field to present research, exchange innovative ideas, propose new models, and demonstrate advanced methodologies and novel systems.

This proceedings is a compendium of selected research papers presented at the 2022 5th International Conference on Geoinformatics and Data Analysis (ICGDA 2022), held virtually, during January 21–23, 2022. Providing broad coverage of recent technology-driven advances, the proceedings is an informative and valuable resource for researchers, practitioners, education leaders, and policy-makers who are involved or interested in Geoinformatics and Data Analysis.

Prestigious experts and professors have been invited as keynote speakers to deliver the latest information in their respective expertise areas. It will be a golden opportunity for students, researchers, and engineers to interact with the experts in their fields of research.

On behalf of the organizing committee, we wish to express our heartfelt appreciation to all the participants and authors. We would like to express our gratitude to the conference committee & program committee members and reviewers. We hope this compilation of papers written by authors worldwide can contribute to the advancement of knowledge, practice, and technology in the emerging Geoinformatics and Data Analysis research.

Warmest regards,

Salah Bourennane
Vit Vozenilek
Conference Chairs

Organization

Conference Committees

Advisory Chair Committee

Witold Pedrycz	University of Alberta, Canada

Conference Chairs

Salah Bourennane	Ecole Centrale Marseille, France
Vit Vozenilek	Palacky University, Czech Republic

Conference Program Chairs

Petr Kubicek	Masaryk University Brno, Czech Republic
Georg Gartner	Vienna University of Technology, Austria
Branislav Bajat	Belgrade University, Serbia
Robert Laurini	Knowledge Systems Institute, France

Conference Program Co-chairs

Caroline Fossati	Ecole Centrale Marseille, France
Mouloud Adel	Université Aix-Marseille, France
Julien Marot	Université Aix-Marseille, France

Conference Publicity Chair

Liu Feng	East China Normal University, China

Technical Committee

Klaus Böhm, i3mainz	Institute for Spatial Information and Surveying Technology Mainz University of Applied Sciences, Germany
Fatima Oulebsir-Boumghar	USTHB, Algeria
Yuh-Jong Hu	National Cheng Chi University, Taiwan
Qiu Chen	Kogakuin University, Japan
Maleerat Maliyaem	King Mongkut's University of Technology North Bangkok, Thailand
Zoran Bojkovic	University of Belgrade, Serbia
Zaher Al Aghbari	University of Sharjah, UAE
Shunbao Liao	Institute of Disaster Prevention, China
Bouchaib Bounabat	ENSIAS - Mohammed V University in Rabat, Morocco
He Zong	Chongqing Geomatics and Remote Sensing Center, China
Tõnu Oja	University of Tartu, Estonia
P. K. Joshi	Jawaharlal Nehru University, India
Lachhman Das Dhomeja	University of Sindh, Pakistan
Oihana Otaegui	Vicomtech, Spain
Konstantinos E. Parsopoulos	University of Ioannina, Greece
Ivan Izonin	Lviv Polytechnic National University, Ukraine
Gokhan Bilgin	Yildiz Technical University, Turkey
Yan Li	Yunnan University, China
Ghani Albaani	Princess Sumaya University for Technology, Jordan
Houda El Bouhissi	University of Bejaia, Algeria
Belkacem Samia	Boumerdes University, Algeria
Helmi Zulhaidi Mohd Shafri	Universiti Putra Malaysia, Malaysia
Pawinee Iamtrakul	Thammasat University, Thailand
Xunhu Zhang	National Quality Inspection and Testing Center for Surveying and Mapping Products, China
Maurizio Leotta	DIBRIS, Università di Genova, Italy
Norbert Pataki	Eötvös Loránd University, Hungary
Francesco Caputo	University of Naples "Federico II" Complesso Universitario di Monte Sant'Angelo, Italy
Wira Hidayat Bin Mohd Saat	Technical University of Malaysia, Melaka
Manik Sharma	DAV University, Jalandhar, India
Hadj Sahraoui Omar	Algerian Space Agency, Algeria
Yi Song	Harbin Institute of Technology Shenzhen, China
Zukun Lu	National University of Defense Technology, China
Liu Lu	Ministry of Land and Resources, China

R. S. Ajin	Kerala State Disaster Management Authority (KSDMA), India
T. N. D. S. Ginige	Universal College Lanka, Sri Lanka
Rushan Abeygunawardana	University of Colombo, Sri Lanka
Ho Ming Kang	Asia Pacific University of Technology and Innovation (APU), Malaysia
T. Arudchelvam	Wayamba University of Sri Lanka, Sri Lanka

Co-sponsored by

Published by

Contents

Geomorphology and Meteorology

Spatio-Temporal Characteristics of Land-Cover Changes in China During 2000–2019

Shu Tao(✉), Tao Cheng, Juan Du, Ran Li, Guangyong Li, and Jin Liu

National Geomatics Center of China, Beijing 100830, China
taoshu@ngcc.cn

Abstract. Land use/cover change is an important theme. In this study, the land cover changes at a national scale during 2000–2019 in China were acquired by the remote sensing and integration of Geographic Information Systems. The spatio-temporal characteristics of land cover changes were revealed by the index of net area of change, single land cover dynamicity model and matrices of land cover change. It can be concluded: (1) affected by policy regulation and economic driving forces, areas with forests, shrubs and grassland may be more easily encroached in recent years. Decision-makers should pay more attention to the conservation of green spaces. (2) Some cultivated land areas with good farming conditions (e.g. abundant rainfall and surface water resources, on the gentle slops) have been occupied for urban expansion. In order to ensure the area of cultivated land being in balance, some places with poor conditions has been reclaimed as compensation for the loss which may aggravate soil erosion or exert pressure on water balance. It is recommended that more attention should also be paid to the quality of cultivated land reclaimed rather than quantity. (3) the artificial surfaces expanded at an accelerated rate with the average yearly increment of 7913 km^2 during the second decade, approximately 1.4 times of the one in the first decade. The increment was distributed particularly in the eastern region and then spread to central and western regions. The result reveals a waste of productive fertile soils. Intensive and economical utilization of urban land may be one way to solve the contradiction between land supply and demand in the economic and social development.

Keywords: Land-cover change · Spatio-temporal distribution · Remote sensing · Land use efficiency

1 Introduction

Land use/cover change (LUCC) has drawn more attention for it is related to human activities on the earth and global environmental change. As the largest developing country, China's economic status has been growing rapidly, which has induced some severe problems during the past three decades such as environmental degradation (Jiang and

S. Bourennane and P. Kubicek (Eds.): ICGDA 2022, LNDECT 143, pp. 3–12, 2022.
https://doi.org/10.1007/978-3-031-08017-3_1

Wang 2016), rampant urban sprawling and encroachment into cultivated land (Kuang 2020). In the 21st century, in order to ensure economic development, grain production and ecological restoration, a series of administrative measures and large-scale programs were implemented by the central government, which drove more rapidly LUCC than ever before in China. The coexistence of various policies potentially forms trade-off relationships and it is usually undetermined these policies were synergistically achieving their respective goals by driving LUCC (Dong 2021). Therefore, studying LUCC is not only an important way for describing regional ecological environments (Liu et al. 2017), but also improving the decision support for land management and ecological restoration. However, previous research has focused mainly on the dynamic monitoring of LUCC at local region of China (Chen et al. 2015; Dong 2021; Liu et al. 2017). Some studies on large spatial scales either usually updated only to 2015 (Liu et al. 2014; Ning et al. 2018) or focused mainly single land cover type (Kuang 2020; Liu et al. 2017).

The objectives of this study were: (1) to reveal the spatiotemporal patterns of China's land-cover change during 2000–2019 by an integrated remote sensing and spatial statistical approaches; (2) to discuss some existed problems unbenifit for the sustainable development and provide a reference for the development of integrated resource management policies.

2 Materials and Methods

2.1 Data Sources

In this research, land cover data with a 5-year interval covering mainland China from 2000 to 2019 were used for analysing the spatio-temporal distribution and conversions of main land cove types. The land cover data in 2000, 2005, and 2010 were generated based on Landsat TM/ETM remote sensing images with a spatial resolution of 30 m and overall accuracy of 86%. The ones in 2015 and 2019 were mostly based on remote sensing images with a spatial resolution of 1 m and overall accuracy of 95%. The land cover types include eight classes, such as cultivated land, forest, shrub, grassland, artificial surfaces, bare land, water bodies, permanent snow and ice (Fig. 1). Moreover, DEM, Precipitation data, statistics on economic and population have also been collected as supplementary data.

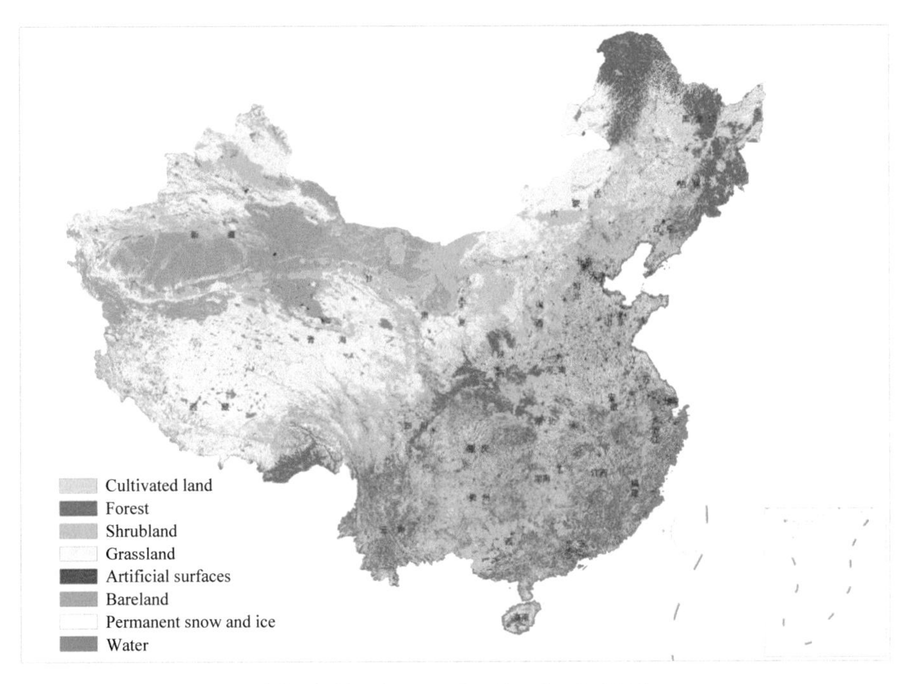

Fig. 1. Land cover classification in 2019

2.2 Methods

(1) In order to evaluate the dynamic land cover changes, net area of change (Ning et al. 2018) and land cover dynamicity model (Liu et al. 2017) have been used to analyze the land cover change rate during the study period.

The single land cover dynamicity model is formulated as follows:

$$\mathrm{K} = \frac{U_b - U_a}{U_a} \times \frac{1}{T} \times 100\% \tag{1}$$

where K is the degree of land cover dynamic, which measures the annual rate change of one land cover type over the given period, and U_a and U_b are the areas of one land type at the beginning (time a) and at the end (time b) of the study period respectively.

(2) In order to evaluate the results of land cover conversion, matrices of land cover change were calculated by a cross-tabulation method and relative changes between years have been determined, which matches the time nodes of various policies.

(3) In order to evaluate the land use efficiency more accurately, two indicators, namely population intensity and economic intensity (econ intensity), have been calculated (Li et al. 2016).

The equation for calculating the population intensity (e_p) was:

$$\mathrm{e}_p = (-1)^n \ln\left|\frac{(p_t - p_0)/p_0}{(A_t - A_0)/A_0}\right|, \quad \begin{cases} p_t > p_0 : n = 0 \\ p_t < p_0 : n = 1 \end{cases} \tag{2}$$

where e_p = population intensity
p_0 = population at the beginning of the period
p_t = population at the end of the period
A_0 = area of land type at the beginning of the period
A_t = area of land type at the end of the period

The equation for calculating the economic intensity (e_g) was:

$$e_g = \ln\left|\frac{(g_t - g_0)/g_0}{(A_t - A_0)/A_0}\right| \quad (3)$$

where e_g = economic intensity
g_0 = GDP at the beginning of the period
g_t = GDP at the end of the period
A_0 = area of land type at the beginning of the period
A_t = area of land type at the end of the period

3 Results and Observations

3.1 Changes in Land Cover Dynamicity

Over the study period, the dominant land cover type in China was grassland covering 28.8% of mainland China, followed by forest (21.6%) and cultivated land (16.8%). Permanet snow and ice area comprised the smallest proportion of land cover type. Between 2000 and 2019, land-cover change across China indicated significant temporal and spatial differences. Net area of changes in land cover categories (LCCs) is presented in Fig. 2. Based on the dynamic model of land-cover change and methods of spatial analysis, the characteristics of main land-cover changes in China has been revealed in Fig. 3.

Artificial Surfaces: The nature of land cover changes revealed that artificial surface categories have been increased significantly. The average yearly increment during the second decade was 7913 km^2, approximately 1.4 times of the one in the first decade. The interpretation of land cover change indicated that earlier expansion was found in the east region, but it started to spread to the central and western regions in the last decade. During 2015 and 2019, an increment in artificial surface of China is 3.94×10^4 km^2, 45% of which is located in the western region, far more than other regions and only 6% in the northeast region. The expansion of artificial surface in the western region was mainly located in flat-terrain areas, such as the Sichuan Basin, Loess Plateau and some areas influenced by the strategies of the Major Function-oriented Zones Planning, while in the central region it was mainly located in Wuhan, Changsha-Zhuzhou-Xiangtan urban agglomerations and other Central Plains.

Cultivated Area: Figure 2 demonstrated that between 2000 and 2015, cultivated land area continued to decrease, but the reduction rate gradually slowed down. The annual reduction decreased from 5200 km^2/year in 2000–2010 to 3400 km^2/year in 2010–2015. The distribution of cultivated area has greatly been constrained by farming conditions

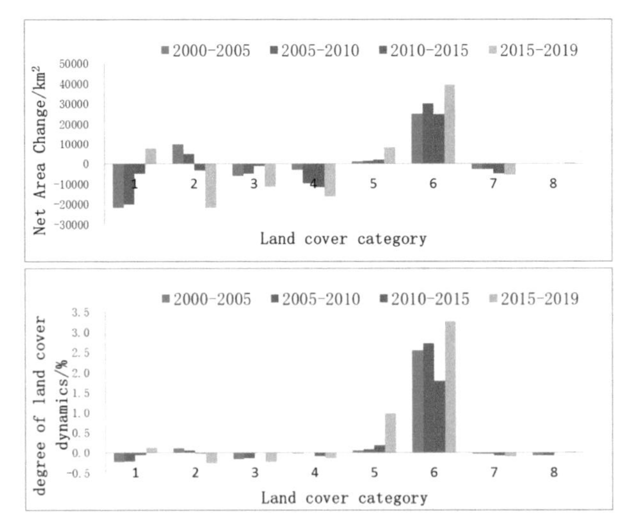

Fig. 2. Net area of change and degree of land cover dynamics in land cover (1: Cultivated land; 2: Forest; 3: Shrubland; 4: Grassland; 5: Water bodies; 6: Artificial surfaces; 7: Bareland; 8: Permanent snow and ice)

such as light, temperature, water, terrain and so on. It has been observed the reclaimed area was mainly in northwest region with poor conditions. For example, from 2015 to 2019, the cultivated land area used for grains decreased by 17.2×10^3 km^2 in the regions with an average annual rainfall over 800 mm, whereas increased by 3.6×10^3 km^2 in the regions with an average annual rainfall less than 800 mm. It can be inferred that it will hardly ensure a sufficient supply of water for crop growth. Moreover, 8.4% of cultivated land reclaimed is located on the hill with slop above 15°, which may aggravate soil erosion and pose a new threat to the ecological environment. In addition, there was 9.6×10^3 km^2 of dry land converted to paddy land for speculative purposes in the east-northern region, which may exert pressure on water balance.

Forest: As the six national forestry projects has been implemented, the forest area increased by 29.8×10^3 km^2, mostly at the expense of cultivated land from 2000 to 2010, which was mainly located in the Loess Plateau and the southern hilly areas. After 2010, the scale and scope of national ecological protection projects have been reduced to a certain extent. Moreover, accelerated farmland reclamation and urban expansion in some areas resulted in the reduction of forest area. The reduction amount and reduction rate have gradually increased. For instance, net area of decrease in forest in China is 22×10^3 km^2 over 2015–2019, in which the conversion of forest to other LCCs was 66×10^3 km^2; other land conversion to forest area was 44×10^3 km^2.

Other LCCs: The area of grassland, shrub and bare land continued to decrease, while water areas continued to increase. In the same period, there is basically no change in the area of permanent snow and ice.

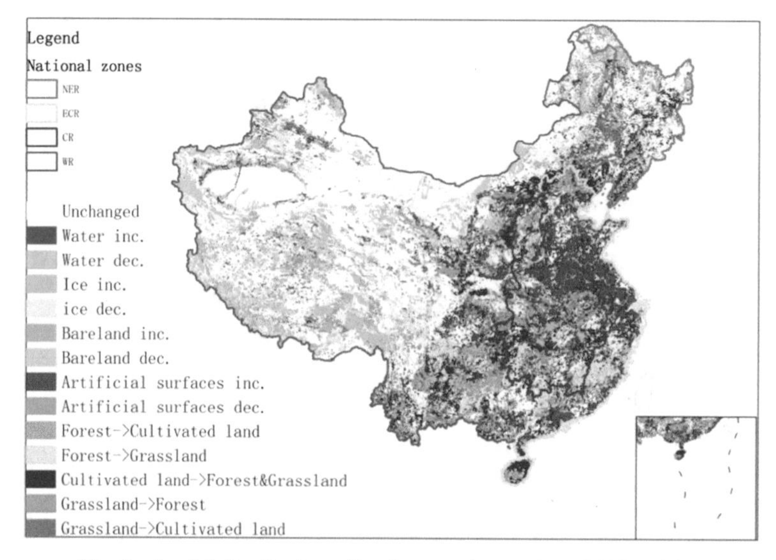

Fig. 3. Spatial distribution of land cover change from 2000 to 2019

3.2 Changes in Land Cover Conversion

Matrices of land cover change from 2000 to 2010, 2010 to 2015, and 2015 to 2019 were calculated and relative changes between years have been determined. It can be learned that the transitions among cultivated, artificial surfaces, forest and grassland were the most significant. Figure 4 shows the distribution of the maximum contributing factors to the main land conversion between 2015 and 2019.

It was concluded that artificial surface was the category having increased significantly in the last decades. Matrices of land cover change revealed that cultivated land contributes mostly to urban development in the 2000–2010, accounting for nearly 77.5% of artificial surface increment. In order to make sure the total area of cultivated land unchanged, more intense protection measures have been taken. Under this circumstance, only 38.1% of artificial surface increment was at the expense of cultivated land from 2015 to 2019, meanwhile more grassland and forest were converted to artificial surface, which separately accounted for 28% and 19% of artificial surface increment. It means less opportunities was left for cultivated lands onto urban expansion, while more vegetation has been occupied to meet up the growing demand of urban land. Moreover, conflicts between artificial surface and other LCCs varies with different regions which can be shown in Fig. 4E. The artificial surface increment was mostly at the expense of cultivated land in north eastern and centre regions of China, whereas forest and grassland in southeast and northwest China.

The change of cultivated land area was one of the categories concerned mostly by the Chinese government. Urbanization contributes mostly to the cultivated land decrease in China in the last 20 years. The area of cultivated land occupied by urbanization accounted for 51.7% during the first decade and 34.3% during the second decade. The conversion among forests, grassland, and cultivated land was another main cause. In order to protect the environment, some national ecological restoration projects have been launched. For instance, the aim of Green for Grain Project is to return farmland to forests and grasslands in the ecologically fragile areas, which is mainly distributed in central and western regions. In the same period, some forests and grasslands especially in the northern arid and semi-arid regions have been reclaimed to make up for the cultivated land occupied. In the first decade, 1.47×10^4 km^2 of grassland and 0.7×10^4 km^2 of forest were reclaimed, respectively accounting for 35% and 17% of the cultivated land increment. From 2015 to 2019, 2.77×10^4 km^2 of grassland and 1.9×10^4 km^2 of forest were reclaimed, respectively accounting for 39% and 27% of the cultivated land increment.

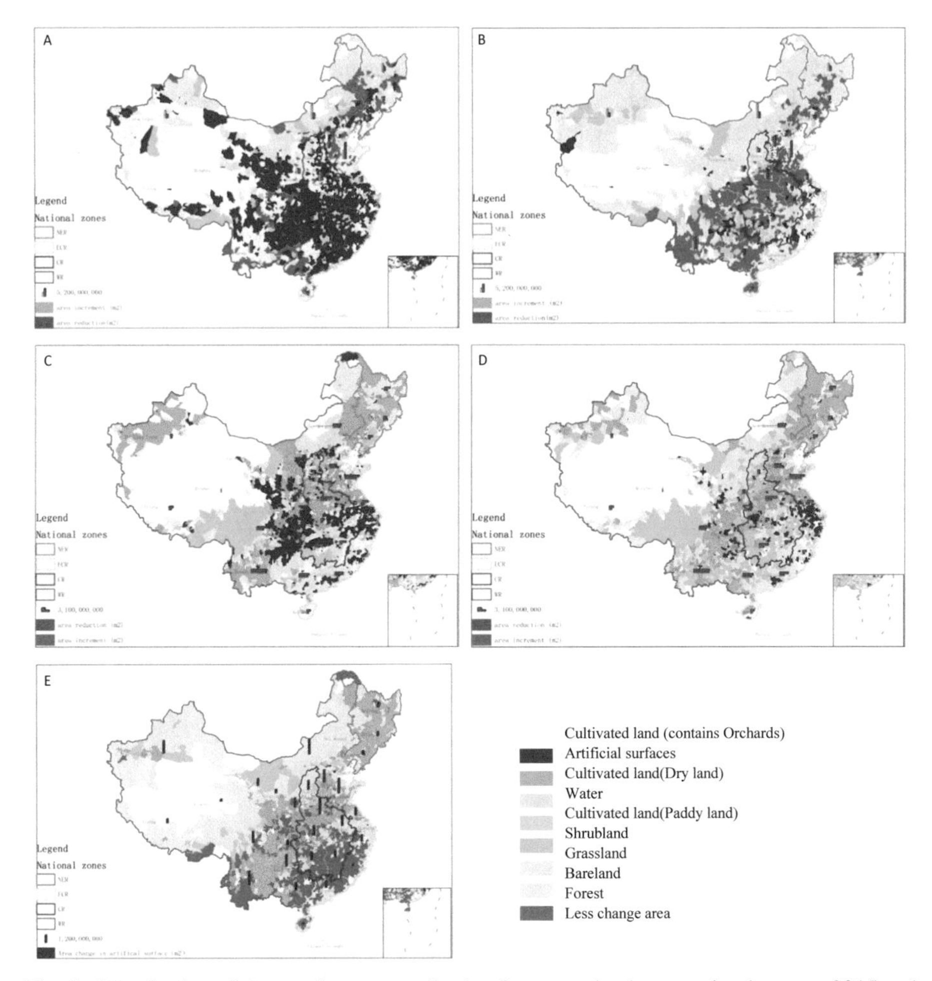

Fig. 4. Distribution of the maximum contributing factors to land conversion between 2015 and 2019 (A: conversion of Cultivated land to other LCCs; B: conversion of other LCCs to Cultivated land; C: conversion of Forest to other LCCs; D: conversion of other LCCs to Forest; E: conversion of other LCCs to Artificial surfaces)

3.3 Land Use Efficiency Across China

As mentioned earlier, the major land-cover change in China was artificial surface expansion in last two decades. Analysis showed that the artificial surface increased by 74% in 2019 compared to 2000. According to the China Statistical Yearbook, the total population of China grew from 1.27 billion in 2000 to 1.4 billion in 2019; GDP increased from 1535.8 to 15271 dollars, with an increase of 9.9 times. In order to illustrate how land cover conversion has been influencing the development of population and GDP, the population intensity and economic intensity in different provinces from 2015 to 2019 have been calculated. Figure 5 shows that the average value of national population intensity is −1.3, indicating that the growth rate of the artificial surface area is more than three times of the national population growth rate. The urbanization rate of land is much faster than that of population, which may reveal a waste of productive fertile soils. This phenomenon is particularly obvious in north eastern and centre regions of China.

The average value of national economic intensity is 1.01, indicating that 1% of artificial surface area increment can only bring about nearly 3% of GDP growth, which is lower than the world's average level. Compared to the provinces with negative GDP growth such as Inner Mongolia and Liaoning, the land use efficiency in the coastal zones is much more intensive.

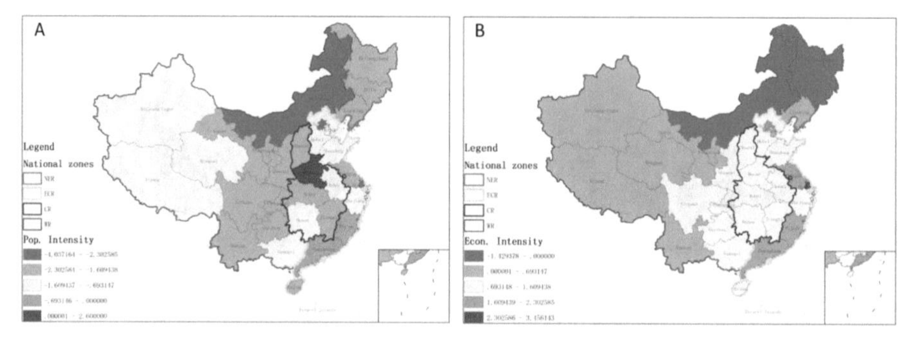

Fig. 5. Distribution of A) population intensity and B) economic intensity between 2015 and 2019

4 Discussion and Conclusions

In response to the continuous decline of environment quality in China, a lot of efforts have been made by the Chinese government to protect the ecosystems (Chen et al. 2019); however, this study revealed that the area of forests, shrubs and grassland may be more easily encroached in recent years caused by rapid urban development and cultivated land reclaim in some regions. It can be even learned that the contradiction between urbanization and cropland conservation may have been shifting to the contradiction between urbanization and ecological protection across China. Therefore, decision-makers should pay more attention to the conservation of green spaces.

The findings of this study also show the existence of some unoptimistic phenomenon related to food security even when the overall area of cultivated land experience a slight increment in the last five years. Firstly, some cultivated land areas with good conditions (e.g. abundant rainfall and surface water resources, on the gentle slopes) have been occupied for urban expansion. Meanwhile other area with poor conditions has been reclaimed as compensation for the loss. In terms of future land conversion programs, we strongly recommend that more attention should also be paid to the quality of cultivated land reclaimed rather than quantity. Moreover, 25900 km^2 of land for crop cultivation was converted to economic crops cultivation in pursuit of higher profit, which was much more than the area (about 10500 km^2) of reversing conversion during 2015 and 2019. It may weaken the functionality of food supply. So it is necessary to optimize the local cultivating structure rather than to expand cropland.

With the ongoing urbanization, the expansion of construction land is inevitable (Chen et al. 2019). However, the urbanization rate of land is much faster than that of population in China, which reveals a waste of productive fertile soils. Intensive and economical utilization of urban land may be one way to solve the contradiction between land supply and demand in the economic and social development. This would also contribute to the ecological equilibrium.

Acknowledgments. The authors gratefully acknowledge the support from Mr. Peng Hou at the Ministry of Ecology and Environmental Protection Cennter for Satellite Application on Ecology and Environment, and Prof. Wenhui Kuang at the Institute of Geographic Sciences and Natural Resources Research, Chinese Academy of Sciences.

References

Chen, H., Marter-Kenyon, J., López-Carr, D., Liang, X.-Y.: Land cover and landscape changes in Shaanxi Province during China's Grain for Green Program (2000–2010). Environ. Monit. Assess. **187**(10), 644 (2015)

Chen, W., Chi, G., Li, J.: The spatial association of ecosystem services with land use and land cover change at the county level in China, 1995–2015. Sci. Total Environ. **669**, 459–470 (2019)

Dong, Y., Ren, Z., Fu, Y., et al.: Land use/cover change and its policy implications in typical agriculture-forest Ecotone of Central Jilin Province, China. Chin. Geograph. Sci. **31**(02), 261–275 (2021)

Liu, J., et al.: Spatiotemporal characteristics, patterns, and causes of land-use changes in China since the late 1980s. J. Geog. Sci. **24**(2), 195–210 (2014)

Jiang, C., Wang, F.: Environmental change in the agro-pastoral transitional zone, Northern China: patterns, drivers, and implications. Int. J. Environ. Res. Pub. Health **13**(2), 165 (2016)

Kuang, W.: 70 years of urban expansion across China: trajectory, pattern, and national policies. Sci. Bull. **65**(23), 1970–1974 (2020)

Li, R., et al.: Spatio-temporal pattern analysis of aritificial surface use efficiency based on Globeland30 (in Chinese). Scientia Sinica Terrae **46**(11), 1436–1445 (2016)

Liu, C., Xu, Y., Sun, P., Huang, A., Zheng, W.: Land use change and its driving forces toward mutual conversion in Zhangjiakou City, a farming-pastoral ecotone in Northern China. Environ. Monit. Assess. **189**(10), 505 (2017)

Liu, W., Liu, J., Kuang, W., Ning, J.: Examining the influence of the implementation of major function-oriented zones on built-up area expansion in China. J. Geog. Sci. **27**(6), 643–660 (2017)

Ning, J., et al.: Spatiotemporal patterns and characteristics of land-use change in China during 2010–2015. J. Geog. Sci. **28**(5), 547–562 (2018). https://doi.org/10.1007/s11442-018-1490-0
Wang, C., et al.: Evaluation of the economic and environmental impact of converting cropland to forest: a case study in Dunhua county, China. J. Environ. Manage. **85**(3), 746–756 (2007)

Analyzing Geospatial Key Factors and Predicting Bike Activity in Hamburg

Cédric Roussel[1](✉), Alexander Rolwes[2], and Klaus Böhm[2]

[1] University of Applied Sciences, 55128 Mainz, Germany
Cedric.roussel@hs-mainz.de
[2] i3mainz, Institute for Spatial Information and Surveying Technology, Mainz University of Applied Sciences, 55128 Mainz, Germany

Abstract. This paper addresses the determination of geospatial key factors, which are relevant for bike sharing stations in the city of Hamburg. They serve as an application case for limited service offers in smart cities. Our approach combines real-world empirical data with open-source data on points of interest for the determination. We apply linear regression methods in combination with an established metric for calculating the geospatial impact. On top of the determination of the geospatial key factors, our paper seeks for machine learning based approaches to predict the bike sharing activity. In our results of the analysis, we identify correlations between bike activity and geospatial factors. Moreover, our neural network provides a solid basis for predicting the activity of bike stations.

Keywords: Geospatial analysis · Predictive analysis · Bike sharing system · Urban analysis

1 Introduction

For the demand-oriented planning or further development of limited service offers, the understanding of geospatial key factors plays a major role. Use cases are for example rental stations for bicycles, parking garages or charging stations. The questions of their location and their capacity are highly dependent on these geospatial factors. For instance, factors are the density or accessibility of shopping facilities, recreational activities, etc. Obviously, not only the geographical proximity, but also the opening hours take a central role. In addition to the planning issues, the prediction of the activity is of great relevance in order to provide the best possible service offer in the operational business. The thematic environment of the paper is bike sharing in the city of Hamburg. Our addressed research questions are:

a. *What are the geospatial key factors affecting the activity of station-based bike sharing in Hamburg?*
b. *How can the activity of mobility use cases be predicted based on the knowledge of geospatial factors?*

We use existing real-world empirical data – booking transactions – to examine these questions.

S. Bourennane and P. Kubicek (Eds.): ICGDA 2022, LNDECT 143, pp. 13–24, 2022.
https://doi.org/10.1007/978-3-031-08017-3_2

2 Related Work

Current research and practice recognize many approaches to identify spatiotemporal relationships between urban (mobility) use cases and geospatial factors. The approaches often form the basis for sustainable urban planning in different sectors. They refer to the occupancy of car sharing or bike sharing systems (BSS). Many approaches use (historical) point of interest (POI) data of *OpenStreetMap*[1] or google places and categorize them into geospatial factors [4, 5, 8, 9, 12, 16]. The reason for this is to simplify the subsequent regression model to avoid overfitting. In contrast, the disadvantage is that the results do not consider individual POI, but rather the categorized geospatial factors. [12] develop a support system for the city of Amsterdam to optimize the locations for electric car charging stations. The approach uses regression models to investigate the impact of geospatial factors on charging stations. [16] investigate the temporal relationships between car sharing and geospatial factors. In addition, they transfer the results of a regression analysis on a further city (Berlin) to predict car-sharing behavior without historical car sharing data of Berlin. The result shows, that the impact of geospatial factors is not the same in all cities. After transferring the model from Amsterdam to Berlin, they point out that 13 of the 20 most influential geospatial factors in Amsterdam differ from those in Berlin. Furthermore, they illustrate the varying impact of geospatial factors over the time of day. Additional research [4, 11, 12] investigates spatiotemporal relationships in different mobility use cases. The research often aims to appoint relocalization systems. Relocalizations are necessary when the pick-up and drop-off of vehicles at stations or in areas is not balanced. Also in BSS, relocalization systems use spatiotemporal relationships in cities to optimize their systems. [7] investigate the location and time demand of the public BSS in Munich based on detailed GPS-data. In addition, this approach utilizes weather data for meteorological correlations and Thiessen polygons to cluster the research area in several quarters. Contrary to many other works, the research of [17] explains, that wind speed is a significant factor in the case of bike sharing as well. They cannot confirm this hypothesis due to the lack of variation in wind strength during the study period. Moreover, BSS represent an important part in urban sustainable planning. [15] implement a geoinformation system (GIS) application to find suitable locations for new bike stations. Citizens and urban planners work together and investigate the geospatial relationships. Based on POI, the application determines distances to transport stations, analyzes demographic data and calculates an index for suitable locations as result. On the contrary, this research does not consider geospatial factors like shops and restaurants, or temporal data. A more intensive consideration of geospatial factors for urban planning takes place in the research of [9]. This research suggests the existence of spatiotemporal relationship between bike sharing and public transit in Cologne. They identify geospatial usage patterns at different times per week by applying negative binomial regression. Besides the geospatial impact of POI on BSS, researchers investigate the reverse relationship to each other. [5] investigate the effect of cycling on local business environments in London. With a two-stage negative binomial regression, they indicate emergence of new local businesses in city quarters, where bicycle traffic

[1] https://www.openstreetmap.de.

is increasing. A similar use case shows the work of [6]. This study determines a significant positive effect of cycling on housing prices in Pittsburgh. A general challenge in dealing with station-based BSS is the decreasing activity during night. The dataset often contains (near) zero values at night times. Therefore, [13] and [17] integrate a zero-inflated-model in their research. [13] determine that future system installation processes should consider the effects of built environment characteristics on bike sharing usage. Based on land use and environment, temporal and weather conditions and public transit facilities in New York, [13] explore a more comprehensive approach to modelling the geospatial and socio-demographic impact on bike sharing in Montreal. They define station-specific regression models using different machine learning algorithms, to identify spatiotemporal relationships.

The main weakness of the existing studies is the lack of consideration of POI opening hours or a weight of each POI in the use case. The existing approach of [8] emphasizes the importance of opening hours in the urban analysis of spatiotemporal relationships. In their research, they present a *metric of geospatial impact* to measure the impact of geospatial factors on off-street parking in Mainz:

$$x_{ij} = \rho_{ij} = \sum\nolimits_{p_k \in P_j} r_{ki} \cdot o_j \cdot \omega_k \tag{1}$$

where

x_{ij} = sum of all relevant values for a particular POI category j at point i
ρ_{ij} = density of POI category j at point i
p_k = specific POI k
P_j = set of POI that belong to category j
r_{ki} = distance decay of a POI k to point i
o_j = opening hour of POI category j
ω_k = weight of POI k in the use case

The metric is the main part of a four-step process to identify geospatial key factors. The equation calculates the density of a defined POI category at a car park garage. It considers the variable r_{ki} of a reachability analysis, the variable o_j for opening hours and the variable ω_k as a weight of an individual POI in the use case. Based on this equation and with the use of geo analytics they identify geospatial key factors.

In our approach, we transfer this process to station-based bike sharing in Hamburg, check it for applicability and optimize the metric of geospatial impact. On this basis, we use the results of the metric to predict the activity of a station with machine learning algorithms. In literature, there is no existing approach to build a prediction model based on geospatial factors with a consideration of opening hours.

3 Use Case and Data

We based our work on booking data of the bicycle rental system *Call a bike*. The data was retrieved from https://data.deutschebahn.com [2]. The operating period is from January 2014 to May 2017. The data contains the date of each trip as well as the GPS-coordinates of the starting and ending station. As research area, we choose the city of Hamburg. The data in Hamburg contains bookings of over 200 bike stations. For each station, we calculate the bookings at every hour in the period of the dataset. We split the data into the four seasons 'Spring' (March, April, May), 'Summer' (June, July, August), 'Fall' (September, October, November), 'Winter' (December, January, February) and into the week intervals 'Weekday' (Monday–Friday), 'Saturday' and 'Sunday'. In addition, we divide the week intervals into the four daytime intervals 'night' (00:00–06:00), 'morning' (06:00–12:00), 'afternoon' (12:00–18:00) and 'evening' (18:00–00:00).

The geospatial data was retrieved using *OpenStreetMap* and contains all relevant POI. In our study, POI are possible destinations for bike users like shopping malls, doctor's offices or movie theaters.

4 Methodology and Data Analysis

This section introduces the methodology to measure the impact of geospatial factors and to predict bookings at stations in Hamburg. For this, we apply the metric of geospatial impact (1) and its respective process. To measure the impact of geospatial factors we use a slot wise multiple linear regression. Furthermore, we use machine-learning algorithms, such as a zero-inflated regression and a neural network, to predict bookings at stations.

4.1 Metric of Geospatial Impact

To identify geospatial key factors, we pass through the following process. First, we define all relevant geospatial factors and categorize them. This is the basis for calculating the geospatial impact by following Eq. (1) for every POI.

Some POI are mapped as an area and not a point. Therefore, we calculate for all POI areas the centroids. We discard irrelevant categories for bike sharing usage such as *car wash*. Overall, there are 56.000 POI in 255 different categories in Hamburg. To reduce the complexity caused by this many categories, we group them in **eleven geospatial factors** according to [1, 4, 9, 10, 12]. The defined categories are *shopping, health, food services, leisure time, grocery, services* and *specialty retail, finance and insurance, education, public sector, religion* and *others*.

The metric defines the three weights o_j, r_{ki} and ω_k. We consider the **opening hours** of a POI, as it cannot affect geospatial factors when it is closed. We calculate the probability that a POI is open by computing the average number of opened POI in this category. The second weight defines how distant a POI is located from a station. We calculate every walking distance from each station to every POI. After that, we do not use isochrones and weight categories like in the used metric. We define the weight using the **distance decay** function from [3]:

$$P(x) = \begin{cases} 1 & x \leq a \\ e^{-ß*(x-a)/1000} & x > a \end{cases} \tag{2}$$

where

$P(x)$ = distance decay

x = distance of a POI to a station
a = threshold where a POI obtains the weight 1
β = flattening of the weight decrease

P(x) represents a weight in the range zero to one. The value x defines the distance between a POI and a station. The variable a describes the distance where a POI obtains the weight one. We set the value to 30 m. Parameter β controls the flattening of the exponential function. For each station, we calculate this parameter individually. The farthest POI, which is still considered relevant, gets the weight 0.1. Farther POI are not considered. We determine the distance of a station to the next neighboring station as the most distant point to consider POI. For example, Fig. 1 shows the graph of the distance decay function and the reachability map when the next neighboring station is 250 m far away. We do not use Thiessen polygons like [14]. There is the possibility that stations are (fully) utilized or empty and users need to use the next neighboring station.

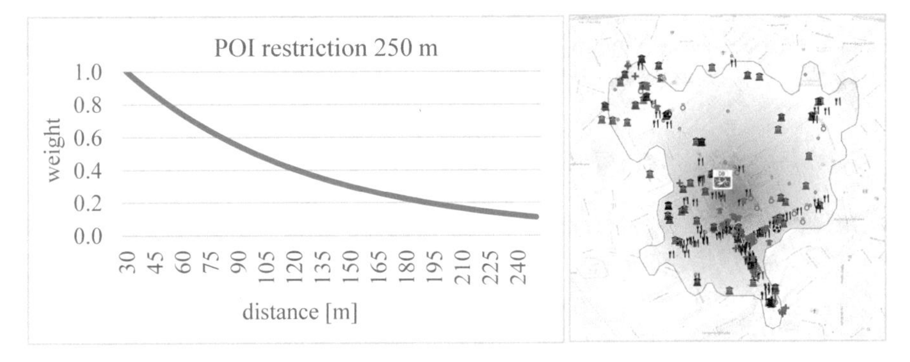

Fig. 1. Distance decay function and reachability map with a restriction to 250 m

To adapt the metric on the use case of bikes, we define the third weight for the **attractiveness** for each of the 255 categories in Hamburg. Therefore, this weight represents the preferences of bike-sharing users. For instance, a shopping mall is considered as more attractive as a veterinary for bike users. Like the opening hours, we define the weight in the range of 0.1 to 1 in ten intervals.

4.2 Key Factors and Prediction

To detect relations between bike usage and geospatial factors, we use a slot wise multiple linear regression analysis. The explanatory variable is the density of each POI, calculated with the metric of geospatial impact. The bookings per hour at a station represents the dependent variable.

The aim of the prediction is to forecast the usage of stations with respect to the geospatial context. For instance, in case of closing POI due to Covid-19 regulations or new stations, where the POI density is different. To get a model with the specific spatiotemporal predictions, we do not divide the data into seasons, weekday intervals or time intervals as mentioned in Sect. 3. Thereby, our model can predict bookings at every month, weekday and hour. In our study we test a zero-inflated regression like [13, 17] and a neural network. A zero-inflated regression performs well in dealing with excess of zeros. In our case, 35% of the data contain the booking number zero. It uses a classifier for the zeros and proceeds with a regressor for the other values. For this model, we split the data into two sections for training and testing. The testing set contains 25% percent of the whole data. After implying the z-score to standardize the data, we set the linear regression as regressor and test different classifier. To evaluate the quality of our model, we measure the score on both sets to eliminate under- and overfitting and consider the root mean squared error (RMSE) on the test set. In our case, the RMSE represents the variance of the predicted bookings.

For the neural network, we use a sequential structure. We set the first layer dimension to the number of attributes in our data. For the last layer dimension, we choose an output of one. To improve our model, we use a training set, validation set and testing set. The testing set is the same size as the validation set. Together they hold 25% of the whole data. We test different hidden layers with activation functions and optimizer to compile the model. For that, we use the keras tuner[2]. We vary the number of epochs and the batch size. We evaluate the quality on the validation set using the score and the RMSE and test the final model on the unknown testing set.

5 Results and Discussion

In this section of our paper, we measure the impact of geospatial factors and evaluate the results. Moreover, we review our machine learning models by testing it on our test data.

5.1 Key Factors

To measure the impact of geospatial factors, we use the multiple linear regression analysis. We evaluate the quality of the regression result by using the score and the p-values at certain significance levels for the coefficients. For the analysis, we reject the three categories *religion, public sector* and *leisure time* with the same reason as [8]. They do not provide enough POI with mapped opening hours. We only consider the categories with more than 10% with mapped opening hours. In the rejected categories, few outliers represent the opening period of an entire category. This indicates spurious correlations.

With four season intervals, three weekday intervals and four daytime intervals we calculate 48 regressions for every combination at every station. Because of very few bookings, the night interval (00:00–06:00) does not show enough significant results. We discard the night interval and consider the other 36 combinations. In general, smaller daytime intervals allow for statements that are more precise. In contrast, smaller intervals result in less data. This leads to less accuracy and fewer significant values. Table 1 and Table 2 show the standardized regression results of two stations in the summer season. We show two results, to display different characteristics in two different locations.

[2] https://keras.io/keras_tuner.

Table 1. Standardized regression results of slot wise multiple linear regression analysis at the station 'Goldbekplatz/Semperstraße' in the summer

		Time	Shopping	Health	Food services	Others	Grocery	Services and specialty retail	Finance and insurance	Education	Adjusted R-squared
Summer	Weekday	06:00–12:00	0.66	2.21***	−0.21	−3.07*	3.14***	0.37	−1.93***	−1.01**	0.66
		12:00–18:00	0.07	−0.14***	−0.21***	0.06	0.22**	0.33***	0.04	−0.81***	0.51
		18:00–00:00	1.26***	0.51*	0.30***	−1.31***	0.28**	−1.33***	−0.31**	1.44***	0.53
	Saturday	06:00–12:00	0.39*	0.09	0.94***	0.01	−0.32***	0.14*	−0.21	−0.26*	0.71
		12:00–18:00	−0.17	0.40	−0.16	0.47	−0.45	−0.53	0.07	0.40	0.09
		18:00–00:00	0.48*	−0.43	−0.58	0.14	0.35	0.16	0.04	0.37	0.27
	Sunday	06:00–12:00	0.23	−0.14	0.47	0.61	−0.14	0.11	0.00	−0.46	0.52
		12:00–18:00	−0.01	−0.02	0.06	−0.08	0.00	0.00	0.00	0.00	0.03
		18:00–00:00	−0.79	0.00	0.46	1.23*	−0.41*	0.26	−0.41*	0.06	0.39

Significance level at 0.001 (***), 0.01 (**), 0.05 (*), n = 4986 observations

Table 2. Standardized regression results of slot wise multiple linear regression analysis at the station 'Mundsburg/Schürbeker Straße in the summer

		Time	Shopping	Health	Food services	Others	Grocery	Services and specialty retail	Finance and insurance	Education	Adjusted R-squared
Summer	Weekday	06:00–12:00	0.71	0.83	−0.26*	−1.00	1.47	0.00	−1.65***	0.04	0.61
		12:00–18:00	−0.09	−0.16***	−0.11	0.27***	0.30***	0.13	−0.08	−0.61***	0.45
		18:00–00:00	1.29***	1.02***	0.37***	−0.70	0.11	−1.42***	−0.29**	0.40	0.52
	Saturday	06:00–12:00	0.08	0.00	1.03***	−0.17	−0.32***	0.21***	0.23	−0.33*	0.62
		12:00–18:00	0.20	−0.33	0.11	−0.46	0.01	0.13	0.28	−0.05	0.06
		18:00–00:00	0.28	−0.09	−0.18	0.19	−0.38	−0.02	0.31	0.38	0.27
	Sunday	06:00–12:00	−0.02	−0.22	0.79*	0.10	−0.22	0.30*	0.00	−0.04	0.51
		12:00–18:00	−0.04	0.11	0.25	−0.11	0.00	0.00	0.00	0.00	0.12
		18:00–00:00	−1.40	0.00	−0.02	1.33	−0.24	1.16	−0.24	−0.15	0.36

Significance level at 0.001 (***), 0.01 (**), 0.05 (*), n = 4986 observations

The station ‘Goldbekplatz/Semperstraße’ is located in the district ‘Winterhude’, between the ‘Außenalster’ and the city park. At this station on weekdays the geospatial factor *health* shows a highly positive significant effect on bike usage in the morning hours ($b = 2.21$; $p < .001$). We explain this by several nearby doctor’s offices, which are open in the morning hours but not in the evenings. Furthermore, the geospatial factor *grocery* has a highly positive significant effect on bike usage in the same time slot ($b = 3.14$; $p < .001$). The impact stays highly positive significant in the weekday and is highly negative significant on Saturday mornings ($b = -0.32$; $p < .001$). We justify it by a close supermarket. It is likely, that workers use bikes to buy things for their working shift. On Saturdays, it is more common that they go by car to do the groceries. The factor *others* shows a significantly positive effect Sunday evening ($b = 1.23$; $p < .05$). We explain this by a dance studio and kiosks, which are open on Sunday evenings. On weekdays the factor *education* has a (highly) significantly negative effect in the mornings ($b = -1.01$; $p < .01$) and at noon ($b = -0.81$; $p < .001$). The station is surrounded by one kindergarten and two elementary schools. The rental bikes usually come at one size, which is too big for children this age to use. Many children in the kindergarten cannot even ride a bike at this young age. Therefore, the effect is significantly negative in the mornings. Evening schools are more likely to be visited by adults. For this reason, this effect turns highly significant positive ($b = 1.44$; $p < .001$) in the evening.

The station ‘Mundsburg/Schürbeker Straße’ is located in the district ‘Uhlenhorst’, to the east of the ‘Außenalster’. The station shows a highly positive significant effect for the category *health* in the evening on weekdays ($b = 1.02$; $p < .001$). We base this on the big medical center ‘Hammonia Bad’, which is opened in the evenings. The factor *shopping* is a key factor in the same time slot ($b = 1.29$; $p < .001$). The station is located near a shopping center with different companies with late opening hours. For instance, the company ‘C&A’, a big clothing company, is open until 20:00. On Saturday ($b = 1.03$; $p < .001$) and Sunday ($b = 0.79$; $p < .05$) morning, the factor *food services* has a (highly) positive significant effect on bike usage. We explain this by nearby coffee shops and restaurants, which provide breakfast.

We cannot compare the impact of the key factors to other research, because the metric is different. [8] used the same metric, but without consider seasons and within another use case.

5.2 Prediction

The following table (see Table 3) shows the results of the zero-inflated regression with different classifier:

The model with decision trees as classifier shows the best result. There is no difference when using 50 or 100 as depth. More depth than 50 does not improve the quality.

Table 3. Results of the zero-inflated regression models

Classifier + linear regression	Trainscore	Testscore	RMSE
Logistic regression (10 iterations)	0.1458	0.1462	4.961
Logistic regression (50 iterations)	0.1455	0.1459	4.962
Decision trees (10 depth)	0.1925	0.1926	4.825
Decision trees (50 depth)	0.2357	0.2254	4.726
Decision trees (100 depth)	0.2357	0.2254	4.726
Random forest (10 trees)	0.2291	0.2189	4.745
Random forest (50 trees)	0.2294	0.2196	4.743
VotingClassifier (Logistic regression, Random forest Decision trees)	0.1988	0.1952	4.816

For the neural network, we implement two hidden layers with the swish activation function. The swish function shows a better generalization on the validation data than other functions. As optimizer, we apply the adam optimizer. In addition, we implement a callback for the adaption of the learning rate after each epoch to reduce fluctuation in the generalization (see Fig. 2):

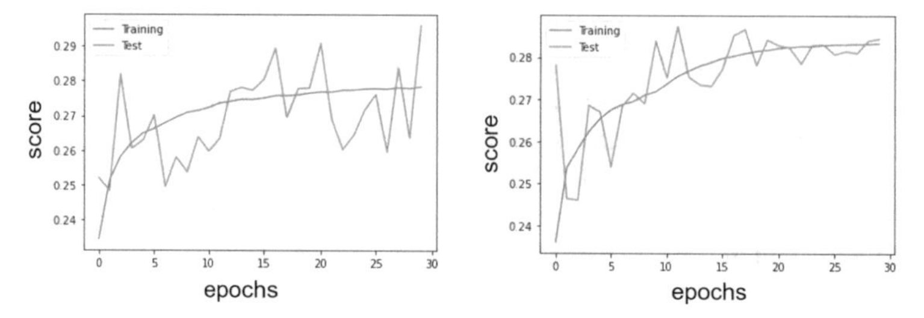

Fig. 2. Fluctuation before and after the callback implementation

We use 30 epochs with a batch size of 200 and train the first ten epochs at normal learning rate. After ten epochs, we reduce the learning rate by using the following formula (3), where the parameter b is set to 0.2:

$$L(x) = x * e^{-b} \tag{3}$$

where

$L(x)$ = new learning rate
x = current learning rate
b = hyperparameter for the decrease

The score on the validation set is 0.28 and the RMSE is 3.2. Therefore, our neural network shows a more precise prediction than the zero-inflated regression. We cannot compare the results of the prediction to current research. We base our research on the metric of geospatial impact, which was not used before.

6 Conclusion and Future Work

In this paper, we analyze bike usage in the city of Hamburg. We measure the impact of geospatial factors on bike usage by using a slot wise multiple linear regression. We identify geospatial key factors on stations in different timeslots. By using a weight for the attractiveness of a POI, the metric can be adapted to other use cases like car sharing. With our weight for the opening hours, we ensure more precise and correct measurements. Especially bigger cities benefit from this approach, as they possess more POI with entries of the opening hour. With our study, we assist growing cities with urban challenges. For instance, the demand for application-oriented analysis options is rising. Our identification of geospatial key factors contributes to sustainable urban planning.

In addition, the visualization of the results is indispensable to make it comprehensible for the domain expert. Statistical tables and figures show the urban relationships between bike sharing activity and geospatial factors. Using geo-visual methods in custom-fit dashboards will help to understand the results by non-statistical experts.

It can be challenging providers to find locations for new stations. Providers need to find locations where stations are required. With our neural network as a machine-learning algorithm, we predict bookings at a new location for a station. Our deviation for predicting the bike activity at every hour in a year is 3.2. Despite the variance, this result can be very helpful for providers, as they often just need an estimated prediction. Furthermore, we can predict the activity on stations with changed POI density. This can help providers in their relocalization strategy. For future research, we recommend training our model with more data from other cities to get more variation in the POI density. This improves the quality of the neural network. In addition, meteorological data could improve the model too. For instance, it is more likely to use bikes on sunny days.

Acknowledgments. The research for this paper is part of the project 'BAM – Big-data-analytics in environmental and structural monitoring' at Mainz University of Applied Sciences. The Carl Zeiss Foundation funds it.

References

1. Bendler, J., Brandt, T., Neumann, D.: An open door may tempt a saint - data analytics for spatial criminology. In: Proceedings of the 24th European Conference on Information Systems, ECIS 2016, Research Paper 98 (2016)
2. Deutsche Bahn AG 2016. Open-Data-Portal – Deutsche Bahn. https://data.deutschebahn.com. Accesed 30 April 2021
3. Gao, K., Yang, Y., Li, A., Qu, X.: Spatial heterogeneity in distance decay of using bike sharing: an empirical large-scale analysis in Shanghai. Proc. Transp. Res. Part D **94**(2021), 102814 (2021)

4. Klemmer, K., Willing, C., Wagner, S., Brandt, T.: Explaining spatio-temporal dynamics in carsharing: a case study of Amsterdam. In: Proceedings of the 22nd Americas Conference on Information Systems, San Diego (2016)
5. Klemmer, K., Brandt, T., Jarvis, S.: Isolating the effect of cycling on local business environments in London. PLoS ONE **13**(12), e0209090 (2018)
6. Pelechrinis, K., Zacharias, C., Kokkodis, M., Lappas, T.: Economic impact and policy implications from urban shared transportation: the case of Pittsburgh's shared bike system. PLoS ONE **12**(8), e0184092 (2017)
7. Reiss, S., Bogenberger, K.: Validation of a relocation strategy for Munich's bike sharing system. In: Proceedings of the International Scientific Conference on Mobility and Transport Transforming Urban Mobility, Munich 2016 (2016)
8. Rolwes, A., Böhm, K.: Analysis and evaluation of geospatial factors in smart cities: a study of off-street parking in Mainz, Germany. In: Proceedings of the 6th International Conference on Smart Data and Smart Cities, Stuttgart 2021 (2021)
9. Schimohr, K., Scheiner, J.: Spatial and temporal analysis of bike-sharing use in Cologne taking into account a public transit disruption. Proc. J. Transp. Geogr. **92**, 103017 (2021)
10. Valizade-Funder, S., Funder, J., Sehi, R.: Situation des innerstädtischen Einzelhandels in Mainz (2018)
11. Wagner, S., Götzinger, M., Neumann, D.: Optimal location of charging stations in smart cities: a points of interest based approach. In: Proceedings of the International Conference on Information Systems, Munich 2013 (2013)
12. Wagner, S., Neumann, D., Brandt, T.: Smart city planning - developing an urban charging infrastructure for electric vehicles. In: 2014 Proceedings of the 22nd European Conference on Information Systems, Tel Aviv (2014)
13. Wang, K., Chen, Y.-J.: Joint analysis of the impacts of built environment on bikeshare station capacity and trip attractions. J. Transp. Geogr. **82**, 102603 (2020)
14. Wang, X., Cheng, Z., Trépanier, M., Sun, L.: Modeling bike-sharing demand using a regression model with spatially varying coefficients. J. Transp. Geogr. **93**, 103059 (2021)
15. Weißmann, K., Pieper, J., Franke, C., Schweikart, J.: Aufbau eines PPGIS zur Standortplanung von Fahrradstationen. J. Appl. Geoinf. **2**, 285–290 (2016)
16. Willing, C., Klemmer, K., Brandt, T., Neumann, D.: Moving in time and space – Location intelligence for carsharing decision support. Decis. Support Syst. **99**, 75–85 (2017)
17. Zhao, De., Ong, G.P., Wang, W., Xiao Jian, H.: Effect of built environment on shared bicycle reallocation: a case study on Nanjing, China. Transp. Res. Part A Policy Pract. **128**, 73–88 (2019). https://doi.org/10.1016/j.tra.2019.07.018

Evaluation and Pattern Optimization of Ecological Space of Chongqing with Remote Sensing Data

Fengmin Wu[1(✉)], Zhipeng Zheng[1], Yalin Wang[1,2], Jian Liu[1], and Shaojia Zhang[1]

[1] Chongqing Geomatics and Remote Sensing Center, Chongqing, China
wufengmin@dl023.net
[2] Southwest University, Chongqing, China

Abstract. Evaluation of ecological space is the basic decision-making for ecological protection and regionalization which closely connected with the variations of ecological functioning. To evaluate the ecological space, we constructed an ecological spatial evaluation system suitable for fragile and sensitive environment by the help of GIS (Geo-Information system) and RS (Remote Sensing). Here we introduced four components for ecological space functioning: ecological sensitivity, ecosystem service value, landscape pattern and ecosystem quality. The ecological sensitivity was comprehensively calculated by soil erosion, rocky desertification, habitat type and acid rain. We measured the ecological service value by integrated method of the value equivalent factor in unit area. Then, we evaluated index of ecological space by judgment matrix method from the above four components. Through overlay of Ecological red line and Nature reserve system, we divided ecological space into high ecological index space, low ecological index space with high sensitivity or service value, low ecological index space and analyzed spatial pattern characteristics of each type. The study provided important references for optimization of land distribution.

Keywords: Evaluation of ecological space · Remote sensing data · Chongqing

1 Introduction

Ecological space is the composition of elements and environment closely related to life phenomena and biological activities which has a certain spatial form, spatial distribution and movement laws [1]. The study of ecological space has particular significance for optimization of ecological pattern and sustainable development, and is the basis for practice of Xi Jinping's Ecological Civilization Thought that lucid waters and lush mountains are invaluable assets [2]. Currently, the research of ecological space evaluation mainly focused on the critical ecological space, such as key ecological space, important eco-function areas, Ecological red line, ecosystem stability and ecological sensitivity [3–9]. Most researchers generally utilized one or several elements of ecological sensitivity, ecosystem service value, and landscape pattern for evaluation of ecological space. There were few studies on comprehensive analysis of all factors in Chongqing which seemed

S. Bourennane and P. Kubicek (Eds.): ICGDA 2022, LNDECT 143, pp. 25–34, 2022.
https://doi.org/10.1007/978-3-031-08017-3_3

very important for differentiate ecological space. In addition, the spatial resolution of remote sensing data has a greater impact on the evaluation results.

The land cover of Chongqing is fragmented and ecosystem seems complex and vulnerable with proportion of mountains and hills more than 90%. The characteristics of ecological space are significantly different from other areas. In the past, the evaluation of ecological space in Chongqing was usually based on single-element research. There were few studies on comprehensive evaluation of multiple elements. Taking Chongqing as the research area, we constructed an ecological spatial system suitable for fragile and sensitive environment including ecological sensitivity, ecosystem service value, landscape pattern and ecosystem quality. The ecological spatial pattern of Chongqing was analyzed by comprehensive assessment of ecological index (see Fig. 1). Where, LAI, NDVI, NPP is the abbreviation of leaf area index, normalized difference vegetation index and net primary productivity, respectively. The delimitation of different ecological space was carried out by overlay analysis of Ecological red line and Nature reserve system.

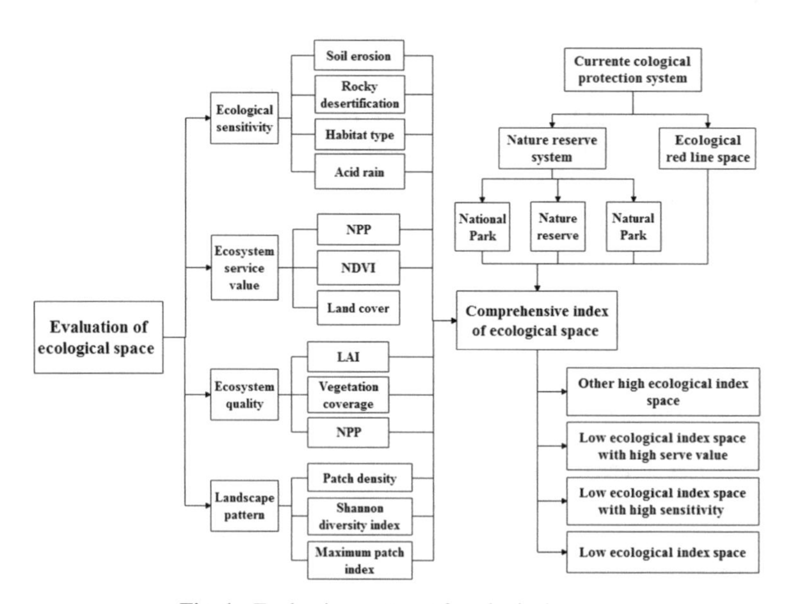

Fig. 1. Evaluation system of ecological space.

2 Methodology

2.1 Data

Chongqing (105°11′E–110°11′, 28°10′–32°13′N) is located in the eastern of the Sichuan Basin of China, with total area of 82.4 thousand km^2. The data used in this study included remote sensing data, meteorological data, soil data, DEM (digital elevation model) data, land cover data and so on. MODIS data was downloaded from National Earth System Science Data Center (https://modis.gsfc.nasa.gov/) to obtain LAI, NDVI and NPP data of 2017. We utilized meteorological data for annual rainfall erosivity factor average monthly rainfall data of 34 weather stations from Chongqing Meteorological Service across the entire period from 2000 to 2014 (http://cq.cma.gov.cn/). Soil data and land cover data were both produced by Southwest University. Soil data included soil texture and soil type. Karst Landform data was from Chongqing Forestry Bureau. Ecological red line and Nature reserve system data were collected from Chongqing Municipal Ecological Environment Bureau. Other data mainly came by Chongqing Geomatics and Remote Sensing Center. The scale of the DEM is 1:5000. All the raster data used in this paper were resampled as 25 m × 25 m and converted to the National Geodetic 2000 coordinate system.

2.2 Ecological Sensitivity

Ecological sensitivity referred to sensitivity to human activities which usually obtained by sensitivity of soil erosion, rocky desertification, habitat types and acid rain comprehensively [10, 11]. We classified ecological sensitivity into 5 levels: non sensitive, less sensitive, medium sensitive, sensitive and extremely sensitive. The soil erosion sensitivity was calculated from the Universal Soil Loss Equation (USLE) as follows:

$$A = R \times K \times LS \times C \quad (1)$$

where, A is the annual average soil loss per unit area, R is rainfall erosivity factor, K is soil texture factor, LS is slope and length factor computed by DEM data, C is vegetation type factor. We got the rainfall erosivity factor calculated as:

$$\mathrm{R} = \sum\nolimits_{\mathrm{i}=1}^{12} (0.3046\mathrm{P_i} - 2.6398) \quad (2)$$

where, P is monthly rainfall. The assignment of each factor on soil erosion was listed in Table1.

Table 1. Sensitivity assessment of soil erosion.

Sensitivity classification	Rainfall erosivity (mm)	Slope and length (m)	Soil texture	Vegetation type
Non sensitive	0–25	0–20	Gravel, sand	Water, herbs
Less sensitive	25–100	20–50	Coarse sand, fine sand, clay	Arbor, shrub, grass
Medium sensitive	100–400	50–100	Surface sand, loam	Shrub, two crops/year
Sensitive	400–600	100–300	Sand loam, silt clay, loam clay	Desert, one crop/year
Extremely sensitive	>600	>300	Sand silt, silt	Non vegetation

Rocky desertification sensitivity was closely related to karst landform, slope and vegetation coverage (Table 2). Karst landform data from Chongqing Forestry Bureau identified the range of karst space of Chongqing. Vegetation coverage was calculated as:

$$f = (NDVI - NDVI_{min})/(NDVI_{max} - NDVI_{min}) \tag{3}$$

where, *NDVI* was obtained from MOD13Q1 products available at pixel resolution of 250 m × 250 m. Sixteen-day composite images were used to eliminate most of cloud cover in daily images. Pixels were moved with reliability score less than zero or greater than one.

Table 2. Sensitivity assessment of rocky desertification.

Sensitivity classification	Karst landform	Slope (°)	Vegetation coverage (%)
Non sensitive	No	<5	≥80
Less sensitive	Yes	5–15	70–80
Medium sensitive	Yes	15–25	50–70
Sensitive	Yes	25–35	20–50
Extremely sensitive	Yes	>35	<20

The acid rain sensitivity could be obtained by soil type, vegetation type, moisture loss. Soil loss equaled rainfall minus maximum evaporation. The habitat sensitivity was calculated through land cover data (Table 3).

Table 3. Sensitivity assessment of habitat type.

Sensitivity classification	Habitat type
Non sensitive	Field crops, orchard, other
Less sensitive	Bamboo forest, economic forest
Medium sensitive	Temperate coniferous forests, shrubs, swamp meadows, aquatic plant
Sensitive	Warm coniferous forest, coniferous and broad-leaved mixed forest, evergreen shrubs, shrubs, typical meadows
Extremely sensitive	Deciduous/evergreen broad-leaved forest, evergreen deciduous broad-leaved mixed forest

Finally, the ecological sensitivity was comprehensively calculated by:

$$ES_j = \sum_{i=1}^{4} W_i f_i \tag{4}$$

where, ES_j is j s unit ecological sensitivity index, W_i is weight factor, f_i is type i factor. The weights of factors of soil erosion, rocky desertification, habitat, and acid rain were assigned value of 0.559, 0.286, 0.116, 0.039.

2.3 Ecosystem Service Value

The ecosystem service value was the natural environmental conditions and ecological processes to satisfy human survival. Quantitative evaluation of ecosystem service value combined with terrain, soil, climate, vegetation and other factors. Total ecosystem service value could be got as follow:

$$V_i = V_{ig} + V_{it} \tag{5}$$

$$V_{ig} = S_i \times H_n \times NPP_i/NPP^{'} \tag{6}$$

$$V_{it} = S_i \times H_n \times NDVI_i/NDVI^{'} \tag{7}$$

$$\mathrm{H}_n = (1/7) \times m\_G \times Ti \tag{8}$$

where, V_i is the total ecosystem service value at pixel i, V_{ig} is the service value of food production and material supplies, V_{it} is the service value of climate regulation, gas regulation, waste treatment, soil conservation, water conservation and biodiversity, H_n is the economic value provided by a certain service function, m_G is ecosystem service equivalent value per unit area obtained from Gaodi Xie's research [12]. *Ti* is standard of equivalent factor economic value, $NPP^{'}$, $NDVI^{'}$ is the average value of NPP, $NDVI$. NPP data was obtained from MOD17A3H of annual average products at spatial resolution of 500 m × 500 m.

2.4 Ecosystem Quality

Sensitivity assessment of ecosystem quality included woodland, grassland and farmland. It was calculated using NPP, LAI and vegetation coverage data. LAI data was from MOD15A2H of eight-day composite images with resolution of 500 m × 500 m. The evaluation indicators of woodland quality included total NPP, relative density of NPP, annual average LAI, annual variation coefficient of LAI, annual average variation coefficient of LAI. Annual average vegetation coverage, annual variation coefficient of vegetation coverage and annual average variation coefficient of vegetation coverage were applied to evaluated grass land quality. Farmland quality was evaluated by annual average NPP, total NPP, annual variation coefficient of NPP and average annual variation coefficient of NPP.

2.5 Landscape Pattern

Ecological landscape pattern reflected the fragmentation, heterogeneity and geometry of the landscape pattern information, quantitatively. In this paper, patch density (PD), Shannon diversity index (SHDI) and maximum patch index (LPI) were taken into analysis. PD expressed the degree of landscape fragmentation calculated by the number of patches per unit area. LPI reflected characteristics of dominant species measured by the ratio of the largest patch area to the total landscape area. SHDI was highly related to landscape heterogeneity calculated as:

$$\mathrm{SHDI} = -\sum P_i \ln P_i \tag{9}$$

where, P_i is the area of each patch type. Through continuous experiment, we found moving window at 500 m × 500 m kept the gradient characteristics without causing large fluctuations in the landscape index. So we used the window size above to calculate SHDI.

3 Analysis

The ecological sensitivity was characterized by high sensitivity in northeast and southeast of Chongqing, and the central and western regions were less sensitive or none sensitive. The distribution of sensitive areas is consistent with the distribution of main mountain system. The extremely sensitive area was roughly in the shape of a 'seven' along Wuxi-the north of Wushan-the south of Fengjie-the southeast of Shizhu-the southeast of Fengdu-the southeast of Wulong. Medium sensitive area was scattered in the northeast, southeast and central mountainous area of Chongqing (Fig. 2).

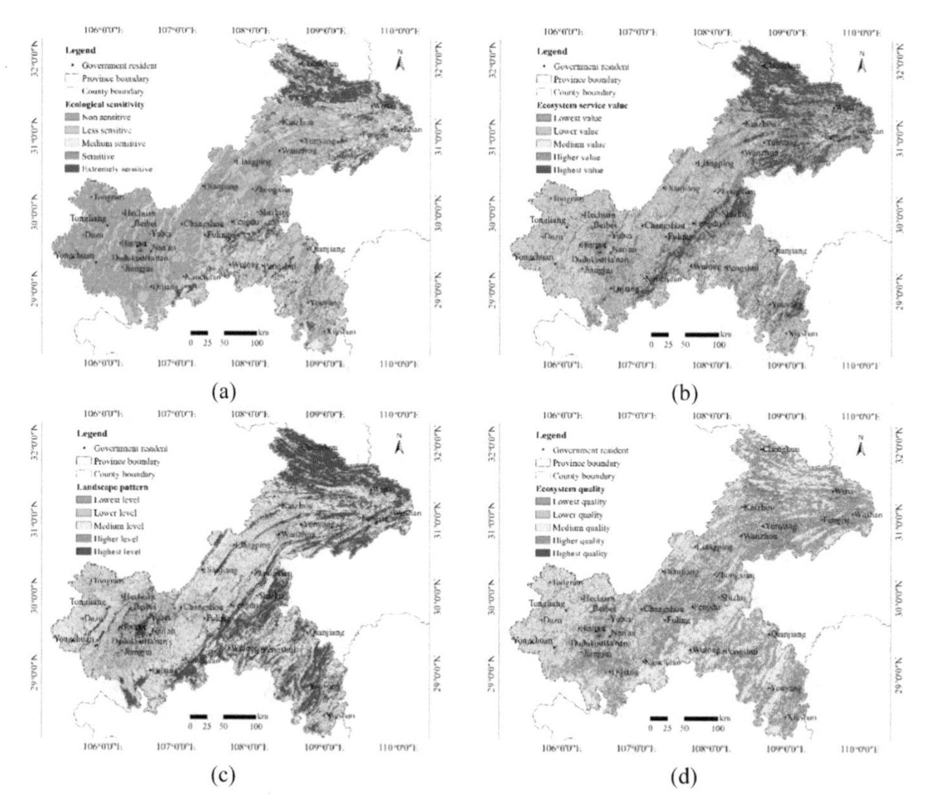

Fig. 2. Distribution of ecological sensitivity (a), ecosystem service value (b), landscape pattern (c) and ecosystem quality (d).

The ecosystem service value had high spatial heterogeneity affected by land cover type, vegetation coverage and topography. The high value areas are mainly distributed in the districts of southeast and northeast Chongqing with high mountains and good vegetation, such as Chengkou, Wushan, Youyang, and Xiushan. At the same time, the ecosystem quality of forest, grassland, and farmland seemed relatively good which mainly concentrated in medium and higher levels. Forest land and farmland were the major types of land cover type and the landscape pattern was better when farther from the urban area.

Based on Natural Breaks (Jenks) method, the four components (ecological sensitivity, service value, quality and landscape pattern) were divided into five grades: lowest, lower, medium, higher, highest reassigned from 1 to 5, separately. Using judgment matrix method to evaluate ecological value grade comprehensively, we finally calculated the evaluation index of ecological space (Fig. 3).

The evaluation index had significant regional differentiation. The area of highest and higher grades covered 30098.14 km^2, accounted for 36.53% of Chongqing. The two grades mainly distributed in northeast (16851.20 km^2) and southeast of Chongqing. There were 7 Districts (Chengkou, Wuxi, Fengjie, Wushan, Wulong, Shizhu and Youyang) with area of highest and higher grades accounted for more than 50% (Table 4). In addition, it was also distributed in the Sishan mountain in Parallel Ridge Valley Area of western Chongqing. The low-grade areas were almost concentrated in urban areas of western Chongqing with low ecosystem service value, low ecosystem quality, or fragmented landscapes.

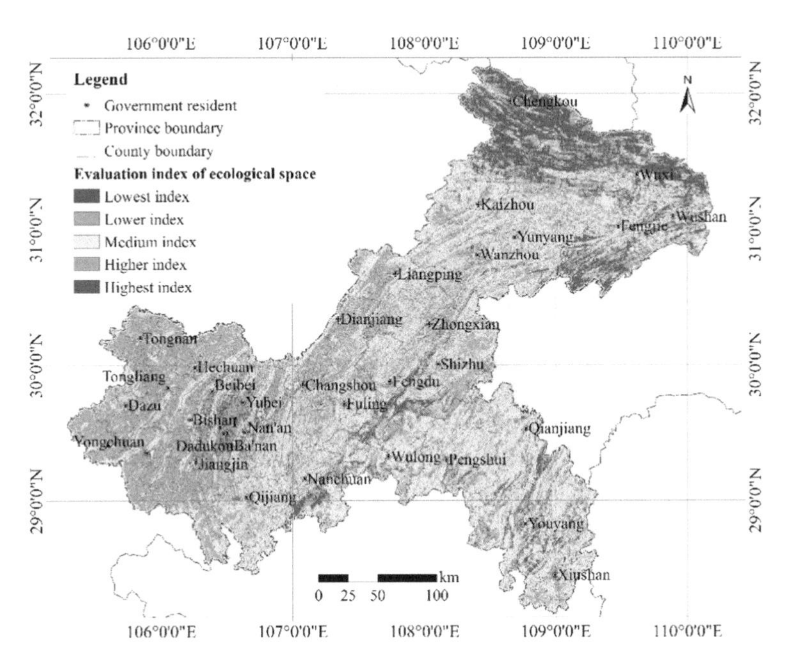

Fig. 3. Evaluation index of ecological space.

Table 4. Area and proportion of Ecological index of ecological space.

Evaluation index	Area (km^2)	Proportion (%)
Lowest	2362.04	2.87
Lower	23286.75	28.26
Medium	26650.99	32.34
Higher	21915.26	26.60
Highest	8182.88	9.93

We divided the ecological space again by overlay analysis of Ecological red line and Nature reserve system. Therefore, the ecological space consisted of five parts: Ecological red line space, Nature reserve system space, other high ecological index space, low ecological index space with high sensitivity or service value, low ecological index space (Fig. 4). For better space management strategies, we combined fragments smaller than 1 km^2 with surrounding large spots.

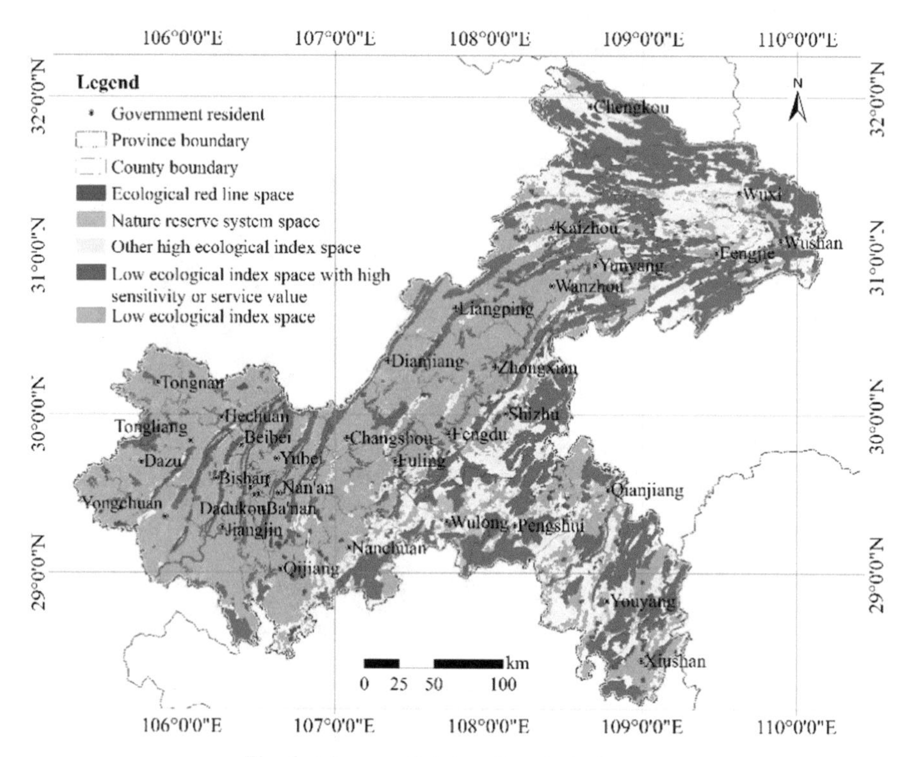

Fig. 4. The new division of ecological space.

The result showed, the area of Ecological red line space was 20445.42 km^2 that accounted for 24.81% of total area of Chongqing. Nature reserve system space outside Ecological red line space covered 5398.42 km^2 which distributed in the neighbourhood of Ecological red line space. Other high ecological index space reached 15486.71 km^2 at percentage of 18.79% mainly distributed in southeast and northeast of Chongqing including Nanchuan, Fuling, Changshou, Yongchuan and Tongliang. Low ecological index space with high sensitivity or service value was totally 11655.68 km^2, of which high sensitivity area was 1116.23 km^2 and high service value area was 10539.45 km^2. High sensitivity area was relatively scattered in Kaizhou, Wuxi, Fengjie, Youyang, Xiushan mainly caused by high rocky desertification area. The rest low ecological index space area covered 29411.73 km^2.

4 Conclusion

In this paper, we constructed an ecological spatial evaluation system of Chongqing by the help of GIS and RS and introduced four components for ecological space functioning: ecosystem service value, ecological sensitivity, landscape pattern and ecosystem quality. Choosing appropriate evaluation indicators to evaluate each factor separately, the evaluation index of ecological space was carried out by judgment matrix method of the four components above. Finally, we divided the ecological space into five parts and analysed the spatial distribution characteristics of different ecological space. The new division of ecological space was carried out by combining with the data of Ecological red line and Nature reserve system. It was obviously that, the new division of ecological space was greatly beneficial to delimit boundaries of production, ecological, and living space.

However, there were some aspects need deepen optimization. The methodology utilized in this paper was almost previous study which need to be improved based on actual application environment. And the resolution of remote sensing data source was almost 500 m $\times$500 m although all data resampled as 25 m $\times$25 m. It was necessary to use high-precision remote sensing data to delimitation of different ecological space accurately.

Acknowledgments. This work is jointly supported by Research and Demonstration on Key Technologies of Urban Government Data Service Integration and Sharing Collaboration of Chongqing, China (Grant No. cstc2018jszx-cyztzxX0037), Research and application of intelligent monitoring and evaluation of geological environment and restoration management of open-pit mines in Chongqing (Grant No. cstc2019jscx-gksbX0042). We are grateful for the data provided by Professor Yuechen L from Southwest University.

References

1. Gause, G.F.: The Struggle for Existence, pp. 134–157. Williams and Wilkins, Baltimore (1934)
2. Tian, H., et al.: Evaluation on the critical ecological space of economic belt of Tianshan north slope. Acta Ecologica Sinica **01**, 1–14 (2020)

3. Yuan, Z., Dazhi, G., Yuanlong, Z., Guizheng, S.: Research on protective planning of urban key ecological space–a case study of Laoshan ecological area in Qingdao. J. Qingdao Univ. Technol. **41**(04), 41–48 (2020)
4. Fangjie, P., Hongzhi, W., Mingjie, S., An, C., Luyao, W.: Research on spatial identification of ecological space in national key ecological function regions based on GIS: a case study of Changyang County in Hubei Province, China. J. Huazhong Norm. Univ. (Nat. Sci.) **54**(04), 658–669 (2020)
5. Ping, Z., Xin, L., Yuhan, Z., Shihao, W., Lin, H.: Tradeoffs and synergies of ecosystem services in key ecological function zones in northern China. Acta Ecologica Sinica **23**, 1–13 (2020)
6. Zhao, X., et al.: Livelihood vulnerability of farmers in key ecological function area under multiple stressors: taking the Yellow River water supply area of Gannan as an example. Acta Ecol. Sin. **40**(20), 7479–7492 (2020)
7. Chuncai, Z., Ye, L., Xiangtao, W., Haitao, M.: The technical method and key issues of the delimitation of ecological protection red line at county scale—a case study of Da'an City, Jilin Province. Territory Nat. Resour. Stud. **01**, 70–74 (2020)
8. Lei, Z., et al.: Evaluation of ecosystem stability assessment of Chinese fir plantations in western hill area around the Sichuan basin based on entropy weight. J. Central South Univ. Forest. Technol. **40**(07), 79–88 (2020)
9. White, H.J., et al.: Quantifying large-scale ecosystem stability with remote sensing data. Remote Sens. Ecol. Conserv. **6**(3), 354–365 (2020)
10. Wang, W., et al.: Distribution of rainfall erosivity R value in China. J. Soil Eros. Soil Conserv. **2**(1), 29–39 (1996)
11. Fu, S.H., Liu, B.Y., Zhou, G.Y., Sun, Z.X., Zhu, X.L.: Calculation tool of topographic factors. Sci. Soil Water Conserv. **13**(05), 105–110 (2015)
12. Xie, G., Zhang, C., Zhang, L., Chen, W., Li, S.: Improvement of the evaluation method for ecosystem service value based on per unit area. J. Nat. Resour. **30**(08), 1243–1254 (2015)

Research on Calculation Model of National Fragile State Based on Climate Vulnerability

Changhua Chen(✉)

Shanghai Open University Fengxian Branch, Shanghai, China
cch0586@126.com

Abstract. The effects of Climate Change will alter the way of to live, and may have the potential to cause the weakening and breakdown of social and governmental structures. Consequently, destabilized governments may result in fragile states. Therefore, a regional instability calculation model is proposed for determining a country's fragility and measures the impact of climate change based on entropy method. Firstly, through the analysis of the factors affecting national instability, the calculation model of national vulnerability evaluation is constructed. In the calculation model, the evaluation index system is composed of 5 s indexes and 26 observing points, i.e. resources and environment, economy, climate, sociality, polity. In the calculation model, the fragile state index can evaluate whether a country is fragile as well as the degree of vulnerable, the climate change index can measure the impact of climate change and the degree of state fragile can be indicated as fragile, vulnerable and stable. An index is used to identify how climate change increases fragility through direct means and indirectly as it influences other factors. Sudan, as one of the top 10 most fragile states, is picked out and the fragile index shows that the country first becomes stable from 1990 to 2010 and then becomes vulnerable. Sierra Leone is chosen to measure the fragility. The fragile index shows that the country becomes vulnerable shortly, and the country will become fragile in 2048. The calculation model is used to analysis the fragile of EU. The fragile index shows that EU is stable.

Keywords: Fragile index · Entropy method · Climate vulnerability index · National vulnerability

1 Introduction

Country's conflicts are enabled by climate change, to include drought, sea level rise, melting glaciers and vary from region to region. These climate change effects will alter the way humans live, reduce government capacity and breakdown of social.

A fragile state is a place where the national government cannot provide basic necessities for its people. As a fragile country, a country has increased the vulnerability of natural disasters, reducing arable land, unpredictable weather and rising temperatures. The country's climate deterioration has led to an increased population pressure conflict.

S. Bourennane and P. Kubicek (Eds.): ICGDA 2022, LNDECT 143, pp. 35–53, 2022.
https://doi.org/10.1007/978-3-031-08017-3_4

Non-sustainable environmental practices, migration and shortage of resources may be the worse state of the country. So the environmental stress does not necessarily lead to conflict, but the government behind will lead to conflict. Now, we will develop a calculation model to solve the following problems.

- Develop a calculation model that determines a country's fragility and meanwhile measures the impact of environment. The calculation model can use of state is fragile, vulnerable, or stable. And to make sure how climates change increases fragility thought directly or indirectly effects.
- From the top 10 most fragile states choose one, and determine how to increased fragility state through climate change. Use this calculation model to show that, without these effects, the state may be less fragile.
- Choose another country that is not on the top 10 lists to use our model to measure its vulnerability and to see when climate change can push it more vulnerable. We define a tipping point and predict when a country reaches the tipping point.
- Our model shows what interventions can mitigate the risk of climate change and prevent a country from becoming a fragile state. Explain the effect of human intervention and predict the total cost of the country's intervention.
- Whether or not our model can use of anywhere. If not, how to perfect our model?

2 Analysis of the Problem

First, modify the abnormal data, normalize the data, and get the weight from Mat lab, so that we can get the contribution of climate change to the regional instability.

2.1 Build the Regional Instability Model Based on Entropy Method

Based on the data of the World Bank [1], the evaluation and evaluation index system is established. The weight of index is determined according to entropy method, so as to get the degree of impact of climate change on the region. Next, we can determine the degree of vulnerability of the country and divide the vulnerability level. Finally, through the causal relationship among the systems, climate change is determined to increase vulnerability by direct or indirect means.

2.2 Climate Change in Sudan and Sierra Leone National Vulnerability

On the basis of the first question, we can determine the impact of climate change in Republic of the Sudan and Sierra Leone on national vulnerability. Based on the data of 1990–2015 year indicators, the evaluation index system is applied to discuss how climate change indicators can increase the vulnerability of the country. According to the weight of each index of climate change, it is determined that the national vulnerability can be reduced or increased by changing those indicators. According to the website data of the World Bank, the critical point of vulnerability is obtained. Based on entropy method, the contribution of each index to the tipping point is determined.

2.3 The Total Cost of Intervention in Climate Change

Taking the self-regulating ability of nature as an example, we analyzed the positive feedback and negative feedback among indicators, so as to determine how to mitigate the impact of climate and determine the total cost of intervention.

2.4 Bring the EU into the Calculation Model

Select the EU into the calculation model to make an analysis of whether the calculation model can be used in smaller 'state' or larger 'states' to improve the model.

3 The Regional Instability Model Based on Entropy Method

3.1 Establish Index System of the Calculation Model

Based on the meaning of vulnerability index, following the principles of comprehensiveness scientificity, and qualitative and quantitative combination of index selection [2, 3], takes 178 countries as the research object, selecting 26 indexes from resources and environment vulnerability, economic vulnerability, climate vulnerability, social vulnerability, political vulnerability these five aspects, so that we can establish a set of fragile state index assessing system. Climate vulnerability is means owing to climate change leads to economy cannot afford. The climate vulnerability mainly including drought, flood, and extreme temperature, cultivated area, cereal yield, and average precipitation. Resources and environment vulnerability mainly including GDP of Unit energy consumption, Per capita renewable's inland freshwater resources, forest area, carbon dioxide emissions, land area. Economic vulnerability primary industry, secondary industry, total fixed capital formation(GFCF), gross domestic savings, external balance on goods and services, gross savings, national income per capita. Social vulnerability mainly including population growth, road traffic death, public health expenditure, passenger volume, population density, enrollment rate and institutions of higher learning secondary school education, total unemployment. Political vulnerability mainly including armed forces military expenditure. See Table 1 in detail.

3.2 Determination of Fragile Index Based on Entropy Method

First, we should make clear the impact of natural disasters, sea level changes, precipitation changes and temperature changes on climate change. Then we use data standardization to deal with the original data. Data standardization is to make the data consistent and clear. The purpose is to eliminate the impact of the difference in dimension and size of the data on the results of the calculation.

Data Standardization

Positive Evaluation Index. The calculation method of positive index is

$$D_{ij} = \frac{x_{ij} - minx_{ij}}{maxx_{ij} - minx_{ij}}, \tag{1}$$

Fragile states index is increased with increasing the positive index.

Negative Evaluation Index

The calculation method of negative index is

$$D_{ij} = \frac{maxx_{ij} - x_{ij}}{maxx_{ij} - minx_{ij}} \quad (2)$$

Fragile states index is decreased with increasing negative index. x_{ij} shows the value of the country i in the index j. And then normalizing the date.

Table 1. The index of vulnerability assessment

The secondary indicators	The three indicators	Unit
Economic vulnerability	Primary industry	%
	Secondary industry	%
	GFCF	USD
	Gross domestic savings	USD
	External balance on goods and services	%
	Gross savings	USD
	National income per capita	USD
Climate vulnerability	Drought, flood & extreme temperature	%
	Cultivated area	*m2*
	Cereal yield	*m2*
	Average precipitation	*mm*
Social vulnerability	Population growth	%
	Road traffic death	%
	Public health expenditure	USD
	Passenger volume	peo
	Population density	km^2/peo
	Enr. ratė, institute. of higher learning	%
	Secondary School Education	%
	Total unemployment	%
Political vulnerability	Armed forces	peo
	Military expenditure	USD
Resources and environment vulnerability	GDP of Unit energy consumption	t/USD
	Per capita renewable's Inland freshwater resources	%
	Forest area	%
	Carbon dioxide emissions	t
	Territory area	hm^2

Normalizing Data. Then, calculate the index value P_{ij}

$$P_{ij} = \frac{D_{ij}}{\sum_{i=1}^{m} D_{ij}} \tag{3}$$

Solving the Weight Based on the Entropy Method. Next we use the entropy method to solving weight. The entropy method can avoid the interference of human factors, make the evaluation result more realities and measure the scale of information. And it can ensure that the set of indicators can reflect the vast majority of the original information [4].

Calculate index entropy and index value Q_j decreased with increasing entropy D_j.

$$Q_j = -\frac{1}{lnm}\sum_{i=1}^{m} P_{ij} lnP_{ij} \tag{4}$$

And m is the number of countries. Because lnm is natural number, so $0 < Q_j < 1$. 2. The weight W_j is calculated as follows

$$W_j = \frac{1 - Q_j}{\sum_{j=1}^{m}(1 - Q_j)} \tag{5}$$

Climate Vulnerability. Climate vulnerability CCi is an important indicator of national vulnerability through climate change. The greater the vulnerability of the climate, the worse the stability of the country, which makes the country more vulnerable. The weight calculated by the entropy method and the proportion of the index value after the standardization of the original data product and sum, we can have climate vulnerability [5].

$$CC_i = \sum_{j=1}^{m} W_j P_{ij} \tag{6}$$

where W_j is the weight of the evaluation index j, P_{ij} is value of the country i under the index j.

Fragile Index. Fragile Index have five indicators. We look for the causality of dynamic variables, such as GDP change and climate change, economic change and social changes [6]. The dynamic regression equation is established by the system dynamics method. The general equation is:

$$CR = f(POL,\ ECO,\ SOC,\ CC,\ RE) \tag{7}$$

where PLO is political fragilely, ECO is economic fragilely, SOC is social fragilely, RE is resources and environment vulnerability.

Because of these index all impact fragile index. So, the fragile index CR_i can be obtained by D_{ij} and W_j as follows:

$$CR_i = \sum_{j=1}^{m} D_{ij} W_{ij} \tag{8}$$

The Contribution of Climate to National Vulnerability. Contribution of the climate vulnerability index (GXD) to the national vulnerability index can be obtained as follows:

$$GXD_i = \frac{CC_i}{CR_i} \tag{9}$$

When the degree of contribution is constant, the national vulnerability index increases with the increase of vulnerability index, and the two are positively correlated.

Combining economic vulnerability, climate vulnerability, social vulnerability, political vulnerability and resources and environment vulnerability data from 178 countries, a country's fragility is divided into 3 levels by the fragile state index. See Table 2 in detail.

Table 2. Vulnerability assessment grade

Vulnerability index	$CR < 4$	$4 \leq CR \leq 5$	$CR > 5$
Vulnerability level	Stable	Vulnerable	Fragile

From Table 2 we can know that a country is fragile when the vulnerability index is less than 5, the country is vulnerable when the vulnerability index is more than 4 and less than 5, and the country is stable when the vulnerability index is less than 4.

From Table 3 we can know that the index system of vulnerability assessment economic vulnerability and social vulnerability account for the largest proportion. In economic vulnerability, the proportion of national income per capita and gross domestic saving is larger. In climate vulnerability, the proportion of average precipitation is larger. In social vulnerability, the proportion of secondary school education, enrollment rate and institutions of higher learning are larger. In resources and environment vulnerability, the proportion of territory area is larger.

Table 3. The weight of the vulnerability assessment index system

The secondary indicators	The three indicators	Weight
	Primary industry	0.013
	Secondary industry	0.026
Economic vulnerability	GFCF	0.066
	Gross domestic savings	0.076
0.31	External balance on goods and services	0.012
	Gross savings	0.027
	National income per capita	0.092
	Drought, flood & extreme temperature	0.010
Climate vulnerability	Cultivated area	0.005

(*continued*)

Table 3. (*continued*)

The secondary indicators	The three indicators	Weight
0.07	Cereal yield	0.014
	Average precipitation	0.042
	Population growth	0.029
	Road traffic death	0.026
	Public health expenditure	0.046
Social vulnerability	Passenger volume	0.046
	Population density	0.049
0.33	Enr. rate, institute. of higher learning	0.054
	Secondary School Education	0.067
	Total unemployment	0.010
Political vulnerability	Armed forces	0.046
0.08	Military expenditure	0.037
	GDP of Unit energy consumption	0.030
Resources and environment vulnerability	Per capita renewable's Inland freshwater resources Forest area	0.050 0.044
0.21	Carbon dioxide emissions	0.005
	Territory area	0.077

3.3 The Regional Instability Model Based on Entropy Method

When taking into account the establishment of a causal relationship between the components of the three indicators. A system contains of 5 subsystems are involved, and the 26 indicators of the 5 subsystems are interconnected and interrelated. For example, the size of cultivated land affects the proportion of the first output value, which affects the economic vulnerability. The average precipitation depth affects the various indicators such as economy, society, resources and environment, to affect many aspects of vulnerability. Therefore, climate can affect the country's vulnerability particularly economic vulnerability by indirect means [7]. See the Fig. 1 in detail.

3.4 Analysis the State Vulnerability Based on Entropy Method

Data Acquisition and Neaten. According to the needs of the purpose of this study, we have selected some necessary information. The data is mainly derived from resources on authoritative websites, such as the World Bank. Based on the above data, the data of five aspects of climates, economies, societies, resources, environment and politics are statistically analyzed. First, the data is preprocessed, including the elimination of abnormal data, the supplementation and normalization of the missing data. The data examined by the World Bank website have the data of some indicators missing and

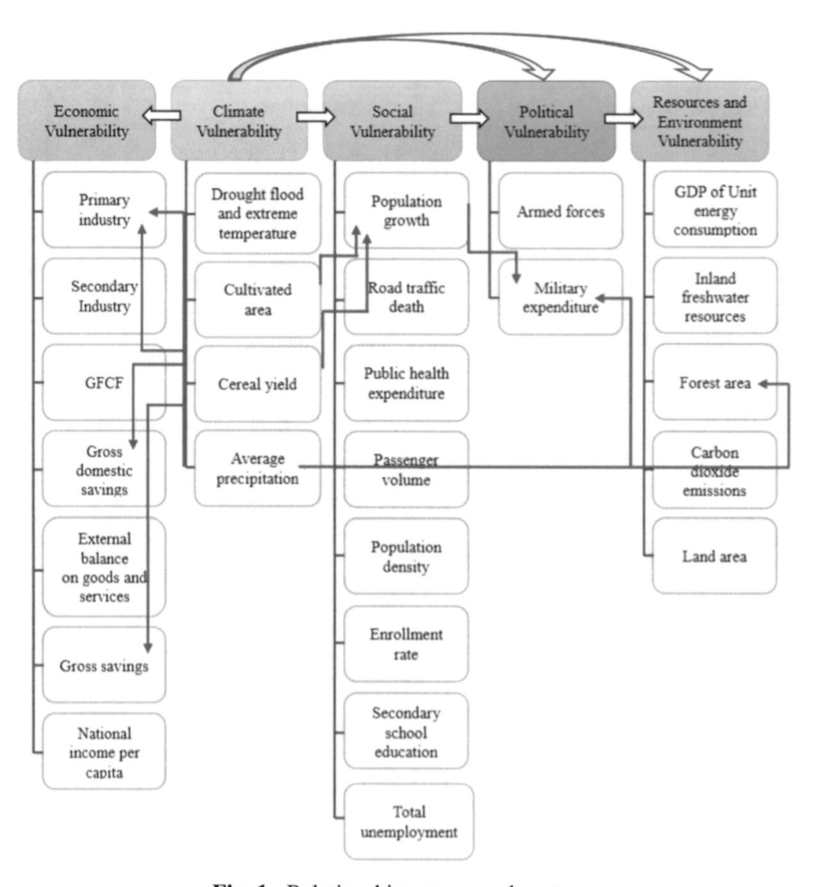

Fig. 1. Relationships among subsystems

abnormal. First, the data of the previous year and the year after the year are analyzed, and the interpolation method is used to fill the missing data. After that, the deviation of the data is corrected. For abnormal data, the "first elimination and post change" principle is used to modify. The data are interpolated according to this principle, and the standard deviation of the variable is similar to that before the interpolation. After the missing and abnormal data processing, the final sample set is obtained.

The Impact of Climate Change on Vulnerability in Public of Sudan Introduce of Sudan. Republic of the Sudan is located in the northeast of Africa, the Red Sea, and the east end of the Sahara desert. Republic of the Sudan is high in 4 weeks and low in the middle. The territory of Republic of the Sudan is vast, and the temperature varies greatly between the north and the south. Agriculture is the main pillar of Republic of the Sudan's economy. The population of agriculture accounts for 80% of the total population of the country. The crops are mainly sorghum, millet, corn and wheat.

After descriptive analysis of the data of Republic of the Sudan climate change, we use entropy method to calculate the weight formula (5) and use Matlab software to get the weight of Republic of the Sudan climate change. According to the index system constructed on Table 1, the index system related to climate change is selected to construct the index system of climate change evaluation. Considering the function of the index, the evaluation index system of climate change is constructed from drought, flood and extreme temperature, cultivated land, grain yield and average precipitation depth, and the index of Republic of the Sudan climate change is determined by improved entropy method [8]. In combination with the formula of contribution degree (9), the impact of climate change on national vulnerability is obtained. See Table 4 in detail.

Table 4. The impact of climate change on vulnerability in Republic of the Sudan

Year	Fragile states index	Climate change index	Climate contribution rate
1990	5.85	0.038	21.9%
1991	5.86	0.037	21.9%
1992	5.82	0.038	21.9%
1993	5.77	0.037	21.6%
1994	5.78	0.038	22.1%
1995	5.83	0.038	22.0%
1996	5.82	0.037	21.5%
1997	5.85	0.037	21.5%
1998	5.82	0.037	21.4%
1999	5.82	0.037	21.5%
2000	5.71	0.037	21.3%
2001	5.74	0.037	21.4%
2002	5.69	0.037	21.3%
2003	5.67	0.038	21.5%
2004	5.56	0.038	21.3%
2005	5.56	0.039	21.6%
2006	5.51	0.039	21.5%
2007	5.45	0.039	21.5%
2008	5.30	0.040	21.0%
2009	5.36	0.040	21.2%
2010	5.28	0.040	21.0%
2011	5.35	0.040	21.2%
2012	5.59	0.038	21.0%
2013	5.68	0.037	21.3%
2014	5.59	0.038	21.1%
2015	5.60	0.038	21.1%

It can be seen from the Table 4 that climate change has more than 20% effects on vulnerability, so climate change has a great impact on vulnerability. Republic of the Sudan has a single economic structure, mainly agricultural and animal husbandry, backward in industry, weak in foundation, and strong in dependence on nature and foreign aid. Republic of the Sudan is one of the hottest countries in the world. The drought and heat are the basic characteristics of the country's climate. In the south, it is hot and rainy in summer, warm and dry in winter, and in the north the high temperature is less rain, the climate is dry and the sand is more windy.

It can be seen from Fig. 2, the climate vulnerability index is related to the national vulnerability index, and the country is most vulnerable when the climate index changes most. So climate change has a direct impact on the vulnerability of Republic of the Sudan.

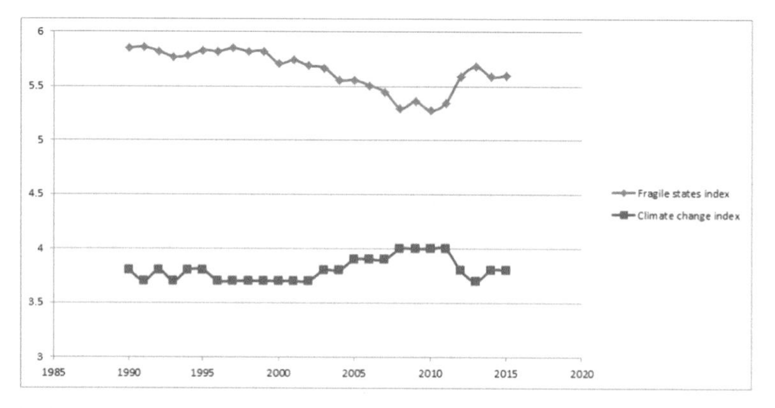

Fig. 2. Fragile states index and Climate change index correlation analysis (Sudan)

As shown in Table 2, the CR > 5 countries is vulnerable and the vulnerability index in the previous table is more than 5. To sum up, climate change directly affects the country's vulnerability and is in line with the actual state of the country, and the calculation model is more reasonable.

The Impact of Climate Change on Vulnerability in Sierra Leone. The Republic of Sierra Leone is located on the Atlantic coast of West Africa. Due to its rich mineral resources, the economy of Sierra Leone is mainly dependent on the mining industry. The Republic of Sierra Leone is a tropical monsoon climate with high temperature and rain. The average annual temperature in the Republic of Sierra Leone is about 26 degrees centigrade. The annual average precipitation is 2000–5000 millimeters, which is one of the most precipitation countries in the West Africa. The low living standards of Sierra Leone were once the source of the supply of European slaves, and were now one of the poorest countries in the world. Whether it is purchasing power, health, longevity or education level is the world's second row, construction is more serious, most of the economic activities are collapse due to civil war.

The Impact of Climate Change on Vulnerability in Sierra Leone. Through the above calculation model, the synthesis of Table 5. As a result of the change in the climate and environment of the Republic of Sierra Leone, the output value of the country's industries has fallen significantly, which leads to the reduction of the domestic total savings. The country is unrest, the rate of school enrolment is reduced, a large number of refugees pour into foreign countries, the population density is reduced, and the state is unable to provide public medical funds and educational funds. The tourism industry has also been greatly impacted, resulting in a decrease in the volume of passenger transport in air transportation. According to statistics, public healthy expenditure, air transport passenger volume, population density and secondary school education shrink to 10%, and total domestic savings become negative. The above indexes raised national vulnerability by 13.6%, 5.9%, 6.9%, 8.8% and 9.7% respectively, and increased the national vulnerability by 44.8% in general.

Table 5. The impact of climate change on vulnerability in Sierra Leone

Year	Fragile states index	Climate change index	Climate contribution rate
1990	3.43	35%	11.9%
1991	3.26	30%	9.8%
1992	3.21	27%	8.6%
1993	3.54	28%	9.9%
1994	3.29	26%	8.7%
1995	3.45	26%	9.1%
1996	3.32	26%	8.5%
1997	3.43	28%	9.6%
1998	3.88	31%	12.2%
1999	3.52	33%	11.5%
2000	3.44	31%	10.6%
2001	3.63	28%	10.1%
2002	3.39	27%	9.0%
2003	3.39	25%	8.3%
2004	3.43	25%	8.7%
2005	3.44	25%	8.5%
2006	3.74	21%	7.7%
2007	3.85	25%	9.5%
2008	4.07	24%	9.8%
2009	4.11	19%	7.9%
2010	3.24	25%	8.0%

(continued)

Table 5. (*continued*)

Year	Fragile states index	Climate change index	Climate contribution rate
2011	3.98	32%	12.6%
2012	3.50	36%	12.6%
2013	5.05	37%	18.8%
2014	3.76	42%	15.7%
2015	4.03	48%	19.4%

The data of Excel was processed to get the curve of vulnerability index (Fig. 3). The results showed that the vulnerability index of Sierra Leone Republic was more obvious and increased year by year.

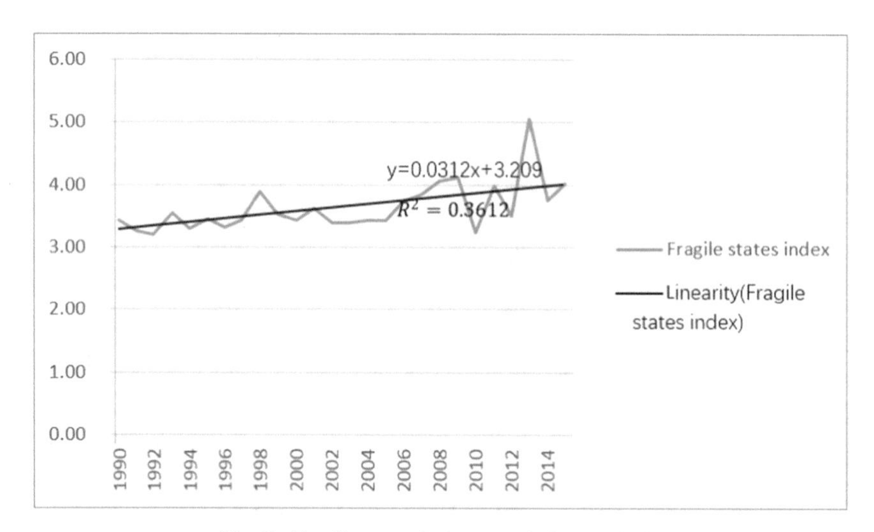

Fig. 3. Fragile states index correlation analysis

The Tipping Point of National Vulnerability. The fragility of the political, economic, climatic, social and resource environment of 178 countries is integrated to eliminate the missing data. Based on the entropy method, the weights of the five vulnerabilities are obtained, and the contribution degree is finally obtained. After enlarging the data, there are about 60 countries' vulnerability index between 4 and 5, 40 countries' vulnerability index is less than 4, and the comparison country vulnerability index table [9], vulnerability index less than 4 is all stable countries, and 4 to 5 are all relatively fragile countries. To sum up, vulnerability can be divided into 3 levels, and when the vulnerability index is less than 5, the country is considered as a fragile state. Therefore, when the vulnerability index is equal to 5, it is a fragile point. See Tables 6 and 7.

Sierra Leone Vulnerability Prediction. Based on determining the above tipping points, we use SPSS to analyze the vulnerability index in time series.

Table 6. National vulnerability based on average value

Country	Vulnerability index	Country	Vulnerability index
United States	1.86	Greece	3.74
Japan	2.34	Portugal	3.80
Norway	2.53	Indonesia	3.87
Canada	2.63	Venezuela	3.91
China	2.69	Peru	3.92
Switzerland	2.78	Saudi Arabia	3.93
Germany	2.85	Chile	3.97
Russia	2.92	Czech Republic	4.03
Australia	3.00	Turkey	4.06
France	3.02	Guyana	4.14
Finland	3.02	Lebanon	4.14
North Korea	3.04	Cyprus	4.18
Austria	3.08	Slovak Republic	4.19
Netherlands	3.14	Philippines	4.19
Italy	3.15	Ecuador	4.19
United Kingdom	3.22	Argentina	4.25
Brazil	3.22	Belarus	4.25
Isr. and West Bank	3.25	Bangladesh	4.26
New Zealand	3.27	Bolivia	4.27
Belgium	3.32	Tri. and Tobago	4.27
Spain	3.35	Mexico	4.32
Ireland	3.50	Poland	4.39
Panama	3.51	Angola	4.42
Colombia	3.51	Algeria	4.44
Indonesia	3.56	Uruguay	4.47
Malaysia	3.65	Honduras	4.48
Hungary	4.51	Namibia	5.45
Thailand	4.52	Kazakhstan	5.48
Belize	4.54	Con. Dem. Republic	5.50
Serbia	4.64	Albania	5.52

(continued)

Table 6. *(continued)*

Country	Vulnerability index	Country	Vulnerability index
Botswana	4.71	Yemen	5.58
Vietnam	4.73	Cameroon	5.58
Paraguay	4.74	Morocco	5.60
Guatemala	4.75	Nigeria	5.67
Romania	4.75	Georgia	5.72
Iran	4.83	Tunisia	5.80
Dom. Republic	4.84	Nepal	5.84
Jamaica	4.86	Pakistan	5.88
Bulgaria	4.88	Cote d'Ivoire	5.89
Ukraine	4.88	Gambia	5.90
Eritrea	4.93	Ghana	5.98
Macedonia	4.99	Armenia	6.02
Zambia	5.07	Senegal	6.04
Azerbaijan	5.10	South Africa	6.06
Nicaragua	5.13	Benin	6.17
Cape Verde	5.16	Moldova	6.19
Cambodia	5.18	Tanzania	6.27

Table 7. National vulnerability based on average value

Country	Vulnerability index	Country	Vulnerability index
Egypt	5.22	Kyrgyz Republic	6.50
Swaziland	5.31	Mozambique	6.91
Guinea Bissau	5.35	Tajikistan	7.30
Mongolia	5.42	Kenya	7.90
Bosnia and Herzegovina	5.44	Togo	7.97

The time series prediction method is an extension and prediction method of historical data, also known as the historical extension prediction method. Time series prediction is a method of extrapolating and predicting its development trend based on the development process and regularity of social economic phenomena reflected by time series (Table 8).

Finally, it predicts that the Republic of Sierra Leone will reach the critical point of vulnerability in 2048.

Table 8. Vulnerability prediction

Year	2016	2017	2018	2019	2020	2021	2022
Vulnerability	4.03	4.06	4.09	4.12	4.16	4.19	4.22
Year	2023	2024	2025	2026	2027	2028	2029
Vulnerability	4.25	4.28	4.31	4.34	4.37	4.40	4.44
Year	2030	2031	2032	2033	2034	2035	2036
Vulnerability	4.47	4.50	4.53	4.56	4.59	4.62	4.65
Year	2037	2038	2039	2040	2041	2042	2043
Vulnerability	4.68	4.71	4.75	4.78	4.81	4.84	4.87
Year	2044	2045	2046	2047	2048	2049	2050
Vulnerability	4.90	4.93	4.96	4.99	5.03	5.06	5.09

4 The Total Cost of Intervention in Climate Change

4.1 Effects of Human Intervention

Effects of Human Intervention. According to Table 3, it is observed that the national per capita net income, the total amount of domestic savings, the total amount of fixed capital formation and the land area have a larger proportion and greater contribution. And the climate change index is increasing year by year, which is due to the joint action of economic, social, political, resource and environmental factors to adjust the vulnerability caused by climate change. The dotted line is negative feedback, and the real line is positive feedback (Fig. 4).

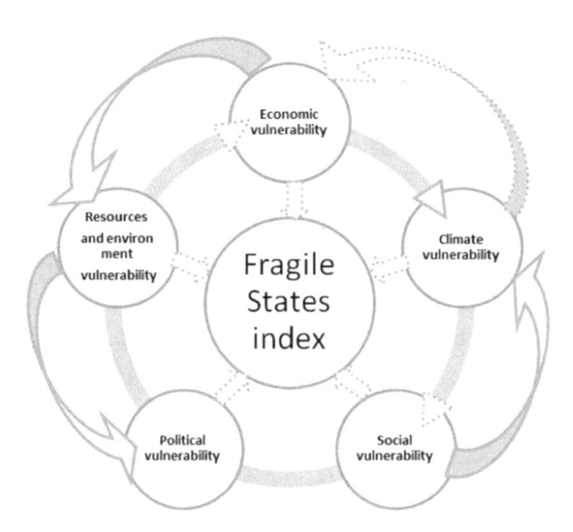

Fig. 4. Influence each other.

Climate change and economic, political, social, and resource and environmental changes are independent and mutual influence. For example, without considering the external state intrusion, when a country in a year of drought and climate change index sharply increased, but the domestic total savings are high, steady state per capita net income, gross fixed capital formation is relatively fixed, large land area, it can make up for the climate change caused by the losses of the country, so that countries vulnerable the index tends to a stable state.

The Total Cost of Intervention in Climate Change. When the climate change is large, under the condition of permission, the government allocations the affected areas, through adjusting the economic changes artificially, reducing the total domestic savings or fixed capital, so as to make up for the harm caused by climate change. At the same time, in the case of the conditions permitted, the forest area will be increased so as to adjust the impact of climate change and ultimately to adjust the state's fragile situation. Due to the large cost of human intervention and the high demand for all aspects of the country, it is not acceptable for vulnerable countries and some relatively vulnerable countries. However, according to the weight of Table 3, increasing the stability of the economy is the fundamental way to reduce the total cost of human intervention.

4.2 The Impact of Climate Change on Vulnerability in European Union

As the EU is a number of countries, each country affects each other but is independent. Most of the EU countries are the Mediterranean climate and the tropical marine climate, which are relatively moist and have a more average monthly precipitation. The climate change in various countries was integrated, and the EU's climate change data were finely tuned. Combined with the calculation model mentioned above, the weight of climate change is solved, thus the fragile contribution of EU climate change to the country is obtained.

Table 9. The impact of climate change on vulnerability in European Union

Year	Fragile states index	Climate change index	Climate contribution rate
1990	3.63	0.066	23.9%
1991	3.74	0.063	23.7%
1992	3.50	0.069	24.2%
1993	3.62	0.052	18.9%
1994	3.34	0.052	17.5%
1995	3.20	0.047	15.2%
1996	3.17	0.045	14.4%
1997	3.05	0.046	14.1%

(*continued*)

Table 9. (*continued*)

Year	Fragile states index	Climate change index	Climate contribution rate
1998	3.02	0.042	12.7%
1999	2.79	0.044	12.2%
2000	2.73	0.046	12.6%
2001	2.52	0.046	11.5%
2002	2.38	0.047	11.2%
2003	2.25	0.040	8.9%
2004	2.18	0.032	6.9%
2005	2.05	0.038	7.9%
2006	1.89	0.040	7.6%
2007	1.76	0.042	7.4%
2008	1.76	0.032	5.6%
2009	1.88	0.036	6.8%
2010	1.88	0.035	6.6%
2011	1.88	0.034	6.4%
2012	1.86	0.036	6.8%
2013	1.77	0.030	5.2%
2014	1.70	0.028	4.8%
2015	1.64	0.036	5.8%

As shown in Table 9, the impact of climate change on vulnerability is decreasing year by year. Most of the European Union countries are moist in the Mediterranean climate and tropical marine climate, and the monthly precipitation is more average and the climate is stable. As shown in Fig. 5, it is clear that the climate vulnerability index is related to the national vulnerability index. Because of the economic, political, resource environment and other factors have great influence on the environment, the increase of the two in the year is decreasing. According to Fig. 5, as a result of the impact of the outbreak of the Libya war in 2011, a large number of refugees poured into the European region, the economic depression in Europe has led to a small increase in the national vulnerability index. And the economic changes indirectly effect climate change, making the climate change index rise slightly.

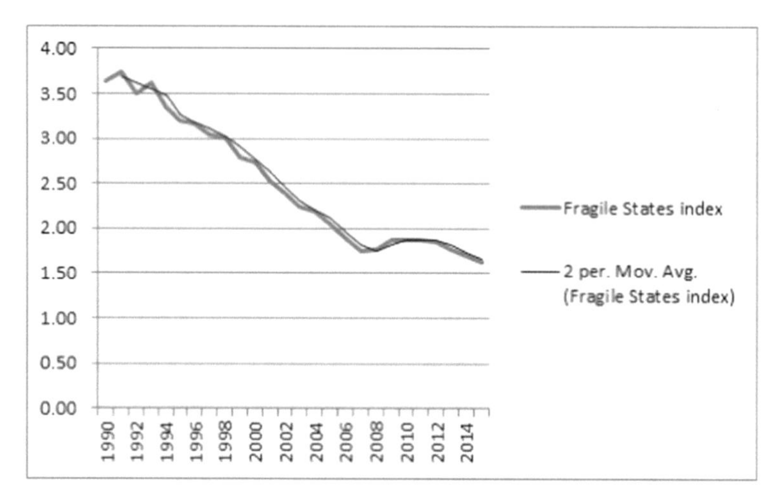

Fig. 5. Fragile states index and Climate change index correlation analysis (European Union)

5 Conclusion

In this calculation model, the concept of the fragile state index was analyzed; then, an indicator system based on the five aspects of economic fragility, political fragility, and social fragility, vulnerability of resources and environment and climate vulnerability. We use standardization to deal with the original data. The purpose is to eliminate the impact of the difference in dimension and size of the data on the results of the calculation. In this paper, the concept of Positive evaluation index is introduced to show the size of the vulnerability index of various countries, and the vulnerability index increases with the increase of the positive evaluation index. This method is simple and easy to operate and has strong operability. Next we use the entropy method of weight to solving weight. The entropy method can avoid the interference of human factors, make the evaluation result more realities and measure the scale of information. And it can ensure that the set of indicators can reflect the vast majority of the original information. It can describe the size and weight of the impact of different factors on the fragile state index. The application process of the proposed method was verified feasible. This computational method was simple and easy to operate. After solving the weight, the national vulnerability index is calculated with 178 national data. Combined with the index weight of climate vulnerability, the contribution of climate change to national vulnerability is obtained. After that, the calculation model is tested in Republic of the Sudan and Mozambique. The results are in accordance with the actual state of the country, and the calculation model is available. Finally, combined with the fine-tuning of climate data of all countries in the European Union, we will bring the fine-tuning data into the calculation model to conduct model checking.

6 Model Evaluation

Strengths:

- On the basis of the hypothesis, a model is established to better measure the vulnerability of the country and simultaneously measures the impact of climate change.
- Using the correct data processing EXCEL method, to make the results more realistic.
- We build a calculation model for influence evaluation of various fields. By the relevant data, we can calculate the climate change to change the state fragility.
- Flexible. The entropy method can avoid the interference of human factors, make the evaluation result more realities and measure the scale of information.

Weaknesses:

- Graphics limited
- Using image processing is only suitable for small quantity of the data. When the data is too large and graphics are difficult to discern, which is optimal.
- Data limit
- Due to less data, we do not know the specific time of some links, so we omit some parameters and simplified calculation model.

References

1. The World Bank. http://www.worldbank.org/en/topic/fragilityconflictviolence/brief/harmonizedlist-of-fragile-situations
2. Guo, X.G.: Entropy method and its application in the comprehensive evaluation, June (1994)
3. Krakowka, A.R., Heimel, N., Galgano, F.: Modeling environmenal security in Sub-Sharan Africa-ProQuest. Geograph. Bull. **53**(1), 21–38 (2012)
4. Xiong, J., Wang, G.W., Deng, C.N.: Comprehensive evaluation of the vulnerability of urban agglomeration in the ring long Zhuzhou Xiangtan City based on entropy weight method, 10-0179-04 (2015)
5. Schwartz, P., Randall, D.: An Abrupt Climate Change Scenario and Its Implications for United States National Security (2003)
6. Theisen, O.M., Gleditsch, N.P., Buhaug, H.: Is climate change a driver of armed conflict. Clim. Change **117**(3), 613–625 (2013)
7. Liu, C.Y., He, W.: Review on the Relationship between Climate Change and Economic Growth (2016)
8. Yan, S.J., Wang, X., Zeng, W.H., Cui, G.N.: Quantitative assessment of climate carrying capacity for cities: a case study of Shanghai City. J. Resour. Ecol. **8**(2), 196–204 (2017)
9. Fragile States Index. http://fundforpeace.org/fsi/

Daily Rainfall Analysis in Indonesia Using ARIMA, Neural Network and LSTM

Syarifah Diana Permai[1](✉) and Ming Kang Ho[2]

[1] Statistics Department, School of Computer Science, Bina Nusantara University, Jakarta, Indonesia
syarifah.permai@binus.ac.id
[2] School of Mathematics, Actuarial and Quantitative Studies (SOMAQS), Asia Pacific University of Technology and Innovation (APU), Technology Park Malaysia, 57000 Kuala Lumpur, Malaysia
dr.ming.kang@apu.edu.my

Abstract. Daily rainfall forecasting is crucial in Indonesia. Because rainfall in Indonesia give rise to floods and landslides, affecting agriculture related to the adequacy of the amount of water on the ground, and affecting transportation, especially sea transportation and air transportation. Daily rainfall data in Indonesia is obtained from Meteorological, Climatological, and Geophysical Agency (BMKG Indonesia). In this research, the daily rainfall modeling was carried out using three methods, which are Autoregressive Integrated Moving Average (ARIMA), Neural Network (NN) and Long Short Term Memory (LSTM). Based on the results of the ARIMA analysis, the model is not good enough because there are no models that meet the assumption of ARIMA. Therefore, rainfall data is analyzed using machine learning methods. The Machine Learning methods that used in this research are NN and LSTM. Based on the results of NN and LSTM, it can be concluded that the NN model is better than the LSTM model. This is due to the root mean square error (RMSE) value on the NN model is smaller than the LSTM model.

Keywords: Daily rainfall · ARIMA · Neural network · Long Short Term Memory

1 Introduction

Indonesia is a country with a tropical climate. In Indonesia, there are two seasons: the rainy season and the dry season. Forecasting rainfall is crucial because floods often occur across the country. In addition, rainfall also affects flights. There are flights that are delayed or cancelled as a result bad weather [1], and airports can also be closed as a result of flood [2]. Therefore, every airport in Indonesia has an observation station, with a total of 92 observation stations [3]. Ngurah Rai Airport is one of the largest international airports in Indonesia which also experiences this incident every year [4]. According to Setiawan (2012) who studied the variability of rainfall and temperature in Bali, Bali is experiencing climate change and rainfall tends to increase [5]. Based on the BMKG

S. Bourennane and P. Kubicek (Eds.): ICGDA 2022, LNDECT 143, pp. 54–65, 2022.
https://doi.org/10.1007/978-3-031-08017-3_5

Ngurah Rai (2020) report, rainfall in most areas in Bali is high to extremely high and above normal [6].

Therefore, in this research, daily rainfall modeling was carried out at Ngurah Rai Airport in Denpasar – Indonesia. The data in this research are time series data. Autoregressive Integrated Moving Average (ARIMA) is the most frequently used statistical method for time series data. However, the ARIMA model has several assumptions that must be satisfied. If the assumptions of the ARIMA model are not satisfied, the model produces inaccurate predictions. There are several methods of machine learning that can be used besides ARIMA model. Machine learning methods that can be used in time series data analysis are Neural Network (NN) and Long Short Term Memory (LSTM). Several studies concluded that machine learning method is better than ARIMA. Besides that, Arima is very strict on assumptions. But ARIMA can perform significant lag testing on the model. Therefore, the comparison between ARIMA and machine learning method is very interesting. As well as the comparison between machine learning method itself.

There are several studies on rainfall analysis using the ARIMA method and machine learning. Ouma, Cheruyot, and Wachera (2021) predicted average monthly rainfall by comparing the LSTM and Wavelet Neural Network (WNN) methods. The research found that LSTM model was better than WNN [7]. Permai, Ohyver, and Aziz (2021) compared ARIMA and NN models to model the daily rainfall. The NN model was shown better than the ARIMA model. In addition, the ARIMA model failed to meet the assumption since the residuals are not normally distributed [8]. Wu et al. (2021) developed a hybrid Wavelet-ARIMA-LSTM (W-AL) model for Total Rainfall and Drought Analysis. The results indicated that Hybrid W-AL model was better than the single ARIMA and LSTM models. However, a comparison of ARIMA and LSTM models revealed that LSTM model was better than ARIMA model [9]. Khan et al. (2020) used LSTM to predict temperature and rainfall in Bangladesh [10]. In addition, the results of the comparison of ARIMA, multiple linear regression (MLR) and Neural Network (NN) on monthly rainfall data in Kirkuk show that the NN model is better than the ARIMA and MLR models [11].

Besides rainfall data, there are several researchers used the ARIMA model and machine learning in analysing a time series data. Ho, Darman, and Musa (2021) predicted the Bursa Malaysia stock price by comparing the ARIMA, NN and LSTM models. The results showed that LSTM model was better than the ARIMA and NN models [12]. The comparison of ARIMA, NN and LSTM models on stock price predictions was also carried out by Ma (2020). Based on the research, it is also concluded that the NN mode was better than ARIMA while LSTM model was better than the NN model [13]. Zhou et al. (2020) compared ARIMA and LSTM on web traffic data and concluded that the LSTM model was better than the ARIMA model [14]. Salman, Heryadi, Abdurahman, and Suparta (2018) who compared the ARIMA, LSTM and LSTM models with intermediate variables. It was found the LSTM model with or without intermediate variables was better than the ARIMA model [15]. Meanwhile, Han (2018) compared the ARIMA and LSTM methods on Australian beer and US beer data. The comparison of the two showed that the ARIMA model was better than the LSTM model. This was because ARIMA can capture seasonal structures in the data, resulting in better predictions than LSTM [16].

Based on this background, there are several researchers who compare ARIMA, NN and LSTM. Some mention NN is better, and some others LSTM is better. It depends on the cases and dataset. Therefore, this research analyzed daily rainfall data in Indonesia using the ARIMA, NN and LSTM methods. This research aims to (1) model daily rainfall data using ARIMA, (2) model daily rainfall data using NN, (3) model daily rainfall data using LSTM, and (4) compare the results of ARIMA, NN and LSTM modeling for obtain the best model using RMSE. Then the best model can help BMKG and Ngurah Rai station for predicting daily rainfall data in Denpasar.

2 Long Short Term Memory (LSTM)

Long Short Term Memory (LSTM) is part and development of the Recurrent Neural Network (RNN). RNN can be used to analyze sequence data, but RNN has a weakness in connecting information from the results of the repetition. Therefore, LSTM was developed to overcome this weakness.

In the LSTM architecture, there are three layers, namely the input layer, hidden layer, and output layer. But in the hidden layer there are memory cells. The memory cell consists of a forget gate, an input gate, and an output gate. Figure 1 is structure of LSTM memory cell [7].

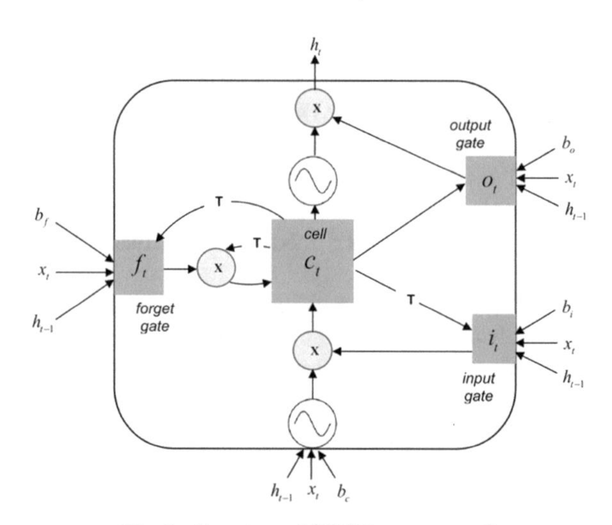

Fig. 1. Structure of LSTM memory cell

2.1 Forget Gate

Forget gates are used to control information stored in memory cells or removed from memory cells. The sigmoid activation function is used in the forget gate. Because the result in the forget gate is the information stored or remove from the memory cell. If $f_t = 0$ then the information will be removed from the memory cell. But, if $f_t = 1$ then the information will be stored in the memory cell. The following is the activation function on the forget gate [12].

$$f_t = \sigma\left(W_f \cdot \left[h_{t-1}, x_t\right] + b_f\right) \tag{1}$$

2.2 Input Gate

The input gate is also used to control the information to be stored in the memory cell. There are two parts to the input gate, namely the input gate layer and the tanh layer. The input gate layer determines the information to be updated. For this purpose, the sigmoid activation function with a value of 0 or 1. The sigmoid activation function is used as follows.

$$i_t = \sigma\left(W_i \cdot \left[h_{t-1}, x_t\right] + b_i\right) \tag{2}$$

Furthermore, the tanh layer updates the value to be stored in the memory cell. The activation function used is the tanh function as follows.

$$\tilde{C}_t = tanh\left(W_c \cdot \left[h_{t-1}, x_t\right] + b_c\right) \tag{3}$$

The value in the previous memory cell will be replaced with the new value. Update values in memory cells is done by processing the values on the forget gate and input gate using the following equation.

$$C_t = f_t \cdot \leq C_{t-1} + i_t \cdot \tilde{C}_t \tag{4}$$

2.3 Output Gate

The output gate determines how many state cells are used to calculate the output. The activation function used is the sigmoid activation function as follows.

$$o_t = \sigma\left(W_o \cdot \left[h_{t-1}, x_t\right] + b_o\right) \tag{5}$$

Then a new cell state value is generated using the tanh activation function as follows.

$$h_t = o_t \cdot tanh(C_t) \tag{6}$$

3 Data and Methods

In this research, the data used is daily rainfall data at Ngurah Rai Airport, Denpasar – Indonesia. This station is a Meteorological Station Class I Ngurah Rai, Denpasar. Data were obtained from the official website of the Meteorological, Climatological, and Geophysical Agency (BMKG Indonesia) [17]. The daily rainfall data used is data from January 1, 2019, to December 31, 2020. There are 731 samples of daily rainfall in this research. The data are time series data. Therefore, rainfall data analysis was carried out using statistical methods and machine learning. Several methods of time series data are used to obtain the best rainfall predictions. Below are the steps to do the analysis:

1. Modeling daily rainfall data using ARIMA. Before modeling the data using the ARIMA, it is necessary to identify the order of p, d and q in the regular or/and seasonal model. After the ARIMA model is conducted, the parameters are estimated and evaluated. If the parameters in the model are not significant, the order of p, d and q must be revised and modeled again. Furthermore, there are two assumptions in the ARIMA modelling. The assumptions are residual must be white noise and normally distributed. Then the ARIMA model that has been obtained must be tested whether the model meets the assumptions or not.
2. Modeling daily rainfall data using NN. There are several things that must be set on the NN, including the number of nodes in the input layer, the number of hidden layers and the number of nodes in the hidden layer, the activation function used from the input layer to the hidden layer, the activation function used in the hidden layer to the output layer, maximum epoch, target error and learning rate. Therefore, to obtain the best NN model, it must be done several times. In this research, NN models are construct from several number of nodes in the input layer and the hidden layer. Then the best model from the NN model is obtained by comparing the RMSE values. In this research, the analysis was carried out using the nnfor package in R. This package is a package for forecasting time series data using neural network, the input layer used is the lags of the dataset [18].
3. Modeling daily rainfall data using LSTM. In LSTM there are also several things that must be set, namely the number of nodes in the input layer, the number of nodes in the hidden layer, and epochs. Therefore, LSTM modeling was carried out several times to obtain the best model. Some models are done by using different number of nodes in the input layer and hidden layer.
4. Comparing the results of ARIMA, NN and LSTM modeling to obtain the best model. This comparison was carried out using RMSE. The RMSE value is calculated for each model that has been constructed. The best model is the model that has the smallest RMSE value.

4 Results and Discussion

Figure 2 is a time series plot of daily rainfall from January 2019 to December 2020. Based on this figure, it can be seen that daily rainfall increases during the rainy season in Indonesia, from October to February. Meanwhile, in the dry season, it can be seen that the daily rainfall can reach 0.

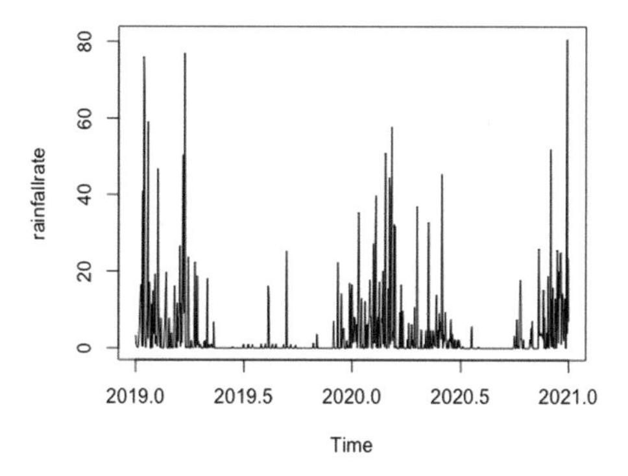

Fig. 2. Time Series Plot of Daily Rainfall.

4.1 Autoregressive Integrated Moving Average (ARIMA) Model

Before constructed the ARIMA model, the stationarity test was carried out on the daily rainfall data. The hypothesis test used is: H_0: Data is not stationary; H_1: Data is stationary.

Based on the Augmented Dickey-Fuller (ADF) test, the statistical value of the Dickey Fuller test is -4.712 and the p-value is 0.01. Because p-value $< \alpha = 5\%$ then H0 is rejected. It can be concluded that the data is stationary then the data can be analyzed using ARIMA and the order of $d = 0$ in the ARIMA model because the data does not need to be differencing.

In the ARIMA(p,d,q) model, it is necessary to identify the ARIMA model by determining the order of p, d and q. Based on the results of the stationarity test, the order of $d = 0$, while the order of p can be determined using PACF plot and order q can be determined using ACF plot. Figure 3 is the ACF and PACF plot for daily rainfall. Based on the identification of ACF and PACF in Fig. 3, there are some significant lags. Therefore, there are several ARIMA models that are constructed. Some ARIMA models which have all significant parameters in the model in the Table 1.

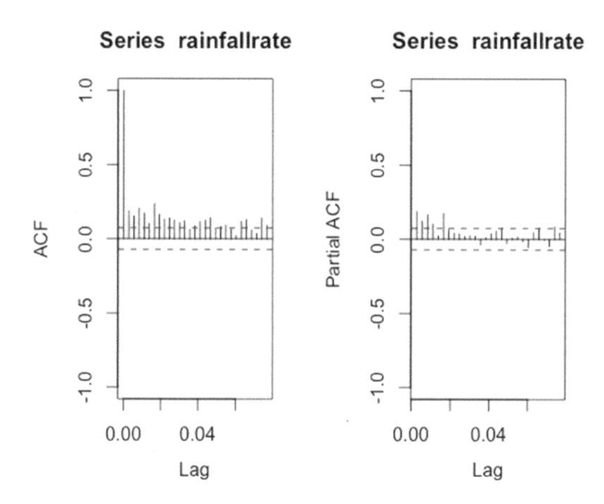

Fig. 3. ACF and PACF plot of Daily Rainfall

Table 1. Comparison of ARIMA Models

Model	RMSE	MAE	AIC
ARIMA(1,0,1)	8.997639	4.330086	5295.414
ARIMA(2,0,0)	9.312424	4.962467	5344.787
ARIMA(3,0,0)	9.184651	4.871841	5326.671
ARIMA(3,0,3)	8.936236	4.531982	5293.267

The comparison of ARIMA model can be seen in Table 1. Based on the comparison of the ARIMA model using the RMSE, MAE and AIC values, it can be concluded that the best ARIMA model is ARIMA(3,0,3). Because the ARIMA(3,0,3) has the smallest RMSE and AIC values than the other models.

Table 2. ARIMA(3,0,3)

Variable	Coefficient	Z test	p-value
AR1	−0.233399	−5.9213	3.194e−09
AR2	0.226720	6.2954	3.066e−10
AR3	0.929489	23.1738	<2.2e−16
MA1	0.346573	7.5246	5.288e−14
MA2	−0.150994	−2.9435	0.003246
MA3	−0.846867	−17.9806	<2.2e−16
Intercept	4.275727	2.8377	0.004545

The parameter estimation of the ARIMA(3,0,3) model can be seen in Table 2. The results of daily rainfall prediction using the ARIMA(3,0,3) model can be seen in Fig. 4. The results of daily rainfall prediction using the ARIMA(3,0,3) can follow the pattern of rainfall data. However, some high levels of daily rainfall cannot be predicted well. It can be seen in the red line that not exceeds 20 while the actual data can reach 80.

The hypothesis testing the significance of the parameters as follows.

H_0: $\beta_i = 0$

H_1: $\beta_i \neq 0$

Based on the results of the parameter significance test using $\alpha = 5\%$, it can be concluded that all parameters in the ARIMA(3,0,3) model have a significant effect on the model. This can be seen from the p-value in Table 2 which is smaller than 0.05. Furthermore, the assumption check on the ARIMA(3,0,3) model is carried out. There are two assumptions that must be met. The assumptions are the residual white noise and the residuals normally distributed.

The hypothesis of L-Jung box test as follows.

H_0: Residuals are white noise

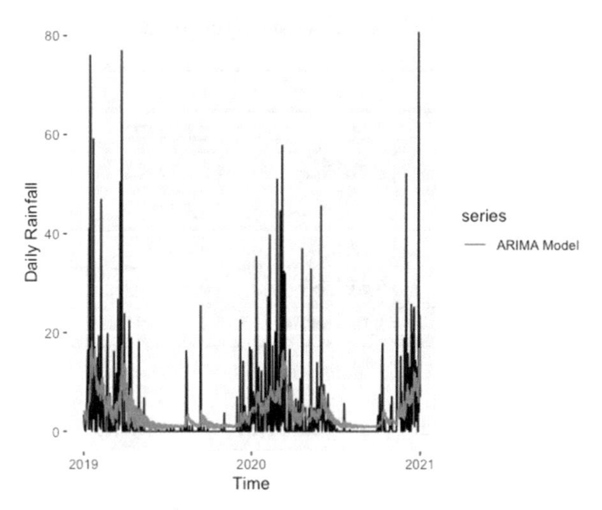

Fig. 4. Daily Rainfall Prediction using ARIMA Model

H_1: Residuals are not white noise

The statistical value of the L-Jung box test is 102.86 and the p-value is 0.9906. Based on these results, the p-value is greater than $\alpha = 5\%$. Then do not reject H_0, it is concluded that the residuals are white noise. This means that the ARIMA assumption is met. Furthermore, the assumption test for the residuals are normally distributed.

The hypothesis of the Kolmogorov Smirnov test as follows.

H_0: Residuals are normally distributed

H_1: Residuals are not normally distributed

The statistical value of the Kolmogorov Smirnov test is 0.30956 and the p-value is $<2.2 \times 10^{-16}$. Based on these results, the p-value is smaller than $\alpha = 5\%$. The H_0 is rejected. It means that the residuals are not normally distributed. It can be concluded that the ARIMA assumption is not met.

4.2 Neural Network (NN) Model

Based on the results of ARIMA model, it can be concluded that the ARIMA model does not meet the assumption that the residuals are normally distributed. Then, an alternative model is needed. One of the models used as an alternative is the Neural Network (NN) model. In the NN model, the inputs that can be used are Y_{t-1}, Y_{t-2}, Y_{t-3}, Y_{t-4} and Y_{t-6}. Based on the PACF identification in Fig. 3, it showed that the significant lags are lags 1, 2, 3, 4 and 6. The several numbers of nodes in the hidden layer was tried to obtain the optimum number of nodes. Table 3 shows a comparison of several Neural Network models.

Table 3. Neural network (NN) models

Model	RMSE	MAE
NN(2,5)	8.90917	4.604322
NN(3,2)	8.706316	4.477288
NN(3,3)	8.653214	4.473696
NN(4,2)	8.661591	4.466811
NN(4,3)	8.597551	4.406792
NN(5,5)	8.008957	4.04797

In Table 3 it can be seen that the smallest RMSE and MAE values are the NN(5,5) model. This shows that the best NN model is NN(5,5). This means that there are 5 nodes in the input layer and 5 nodes in the hidden layer. The architecture of the NN(5,5) model can be seen in Fig. 5.

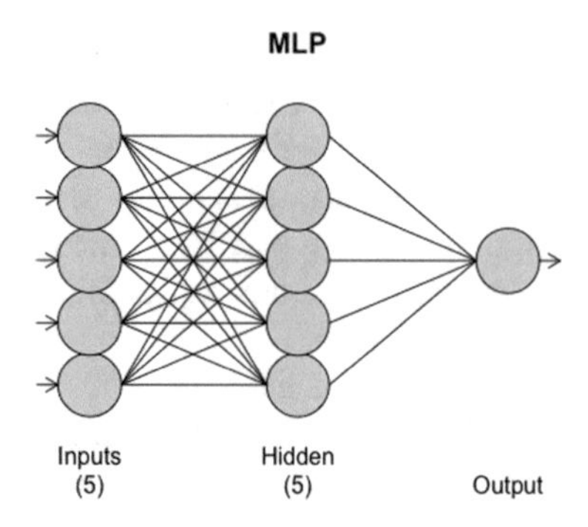

Fig. 5. The architecture of NN model

Figure 6 is a daily rainfall prediction using NN(5,5) model. In Fig. 5. It can be seen that the prediction of NN(5,5) model can follow the actual data pattern of daily rainfall. Moreover, some high daily rainfall levels can also be predicted well. This can be seen in the red line that follows the actual data pattern and some data with high rainfall are also predicted with high values.

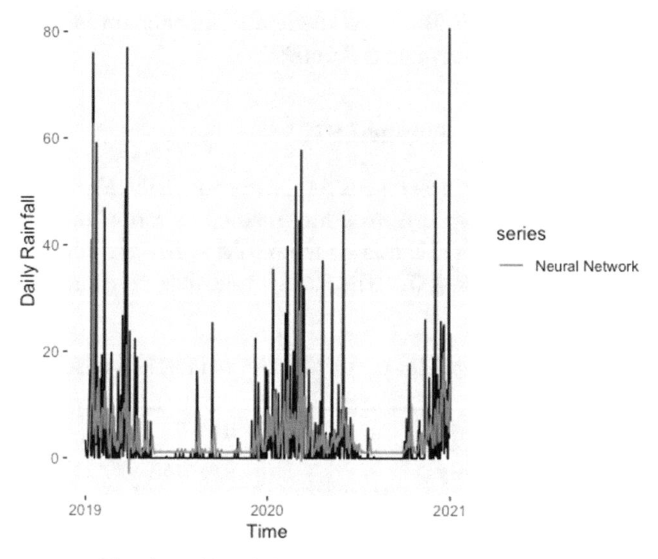

Fig. 6. Daily rainfall prediction using NN model

4.3 Long Short Term Memory (LSTM) Model

The LSTM model uses 5 inputs and 3 hidden layers. While the epoch used is 100. The activation function used is Rectified Linear (ReLu) and the optimization algorithm used is Adaptive moment estimation (Adam). The results of daily rainfall prediction using LSTM model with 3 hidden layers can be seen in Fig. 7.

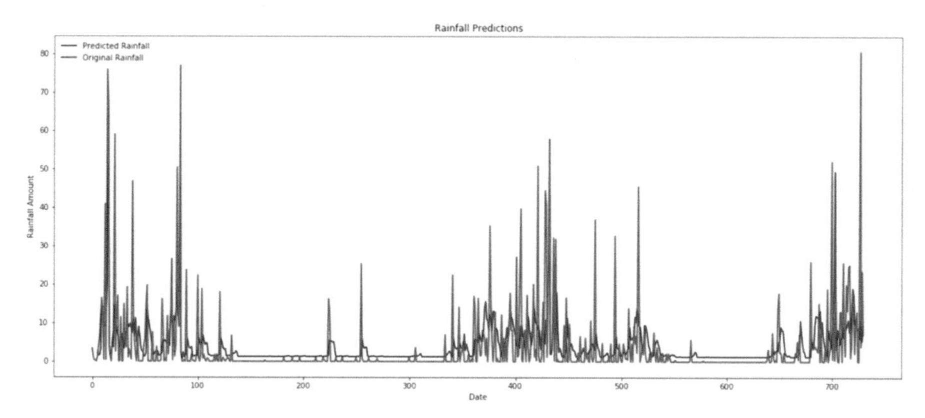

Fig. 7. Daily rainfall prediction using LSTM model.

Based on Fig. 7 the daily rainfall prediction using the LSTM model with 3 hidden layers can follow the rainfall data pattern. However, some high levels of daily rainfall are predictable, and some are not. This can be seen in the blue line following the red line pattern. However, data with high rainfall can only be predicted up to 50. Meanwhile, high rainfall data can reach 80. Based on the RMSE calculation in the LSTM model with 3 hidden layers, the RMSE value is 8.090997.

4.4 Comparison of ARIMA, NN and LSTM Models

This research aims to obtain the best results for forecasting daily rainfall in Indonesia. To achieve this objective then a comparison of three methods is carried out. The comparison of daily rainfall predictions in Indonesia on the ARIMA, NN and LSTM models using the Root Mean Square Error (RMSE). The RMSE value can be seen in Table 4 below.

Table 4. Comparison of ARIMA, NN and LSTM models

Model	RMSE
ARIMA(3,0,3)	8.936236
NN(5,5)	8.008957
LSTM with 3 hidden layers	8.090997

The best model is the model that has the smallest RMSE value. Table 4 shows the comparison of RMSE values in the ARIMA, NN and LSTM models. Based on the result, it showed that RMSE of NN(5,5) is 8.008957. Actually, it is not too different from RMSE value of LSTM with 3 hidden layers, because the RMSE value is 8.090997. Thus, the smallest RMSE value is the NN model (5,5). It can be concluded that the best model is the NN model.

5 Conclusion

In this research, daily rainfall predictions in Indonesia were made from January 1, 2019 to December 31, 2020. This daily rainfall analysis is a time series forecasting. ARIMA is one of the most frequently used time series methods. Based on the ARIMA model, the best model is ARIMA(3,0,3). All parameters in the ARIMA model are significant in the model. However, the ARIMA(3,0,3) model only fulfills the assumption of residual white noise and does not meet the assumption that the residuals are normally distributed. Therefore, daily rainfall modeling was carried out using the Neural Network and LSTM methods. Based on the NN model, the best model is NN(5.5). This model showed that there are 5 nodes in the input layer and 5 nodes in the hidden layer. While in the LSTM model there are 5 nodes in the input layer and 3 nodes in the hidden layer. Furthermore, a comparison of daily rainfall predictions on the ARIMA, NN and LSTM models was carried out using the RMSE value. Based on the RMSE value, it can be concluded that the NN model is the best model because it has the smallest RMSE value than other models. For the further research, NN and LSTM modeling can be carried out using different preprocessing data, for example by handling missing values or outliers. besides that, it can also do time series modeling with other methods such as deep learning.

Acknowledgments. This work is supported by Research and Technology Transfer Office, Bina Nusantara University as a part of Bina Nusantara University's International Research Grant entitled Rainfall Modeling to Prevent Flooding in Jakarta using Machine Learning Method with contract number: No. 026/VR.RTT/IV/2020 and contract date: 6 April 2020.

References

1. Dinisari, M.C.: Ekonomi Bisnis, 16 July 2016. https://ekonomi.bisnis.com/read/20160716/98/566451/penerbangan-domestik-di-bali-terganggu-cuaca-buruk. Accessed 28 Aug 2021
2. Setiawan, B.: Nasional Tempo, Tempo, 3 February 2016. https://nasional.tempo.co/read/741852/hujan-deras-guyur-bali-bandara-ngurah-rai-sempat-ditutup. Accessed 28 Aug 2021
3. BMKG. Informasi Cuaca Aktual Bandara. Badan Meteorologi, Klimatologi dan Geofisika (BMKG), 3 September 2021. https://www.bmkg.go.id/cuaca/cuaca-aktual-bandara.bmkg. Accessed 3 Sept 2021
4. NV. Delapan Penerbangan Luar Negeri ke Bali Cancel, Nusa Bali, 2 December 2017. https://www.nusabali.com/berita/21894/delapan-penerbangan-luar-negeri-ke-bali-cancel. Accessed 3 Sept 2021
5. Setiawan, O.: Media Neliti, 22 April 2012. https://media.neliti.com/media/publications/95390-ID-analisis-variabilitas-curah-hujan-dan-su.pdf. Accessed 6 Sept 2021
6. Rai, B.N.: Dampak La Lina, BMKG Ngurah Rai, Denpasar (2020)
7. Ouma, Y.O., Cheruyot, R., Wachera, A.N.: Rainfall and runoff time-series trend analysis using LSTM recurrent neural network and wavelet neural network with satellite-based meteorological data: case study of Nzoia hydrologic basin. Compl. Intell. Syst. **8**(1), 213–236 (2021)
8. Permai, S.D., Ohyver, M., Aziz, M.K.B.M.: Daily rainfall modeling using Neural Network. J. Phys. Conf. Seri. **1988**(1), 012040 (2021)
9. Wu, X., et al.: The development of a hybrid wavelet-ARIMA-LSTM model for precipitation amounts and drought analysis. Atmosphere **12**(1), 74 (2021)
10. Khan, M.M.R., Siddique, M.A.B., Sakib, S., Aziz, A., Tasawar, I.K., Hossain, Z.: Prediction of Temperature and Rainfall in Bangladesh using Long Short Term Memory Recurrent Neural Networks. IEEE, Turkey (2020)
11. MuttalebAlhashimi, S.A.: Prediction of monthly rainfall in Kirkuk using artificial neural network and time series models. J. Eng. Developm. **18**(1), 129–143 (2014)
12. Ho, M.K., Darman, H., Musa, S.: Stock price prediction using ARIMA, neural network and LSTM models. IOP J. Phys.: Conf. Ser. Kuantan (2021)
13. Ma, Q.: Comparison of ARIMA, ANN and LSTM for stock price prediction. E3S Web Conf. **218**, 01026 (2020)
14. Zhou, K., Wang, W.Y., Hu, T., Wu, C.H.: Comparison of time series forecasting based on statistical ARIMA Model and LSTM with attention mechanism. J. Phys.: Conf. Ser. **1631**(1), 012141 (2020)
15. Salman, A.G., Heryadi, Y., Abdurahman, E., Suparta, W.: Weather forecasting using merged long short-term memory model (LSTM) and autoregressive integrated moving average (ARIMA) model. J. Comput. Sci. **14**(7), 930–938 (2018)
16. Han, J.H.: Comparing Models for Time Series Analysis. Wharton Research Scholars, Philadelphia (2018)
17. BMKG. Data Online Pusat Database BMKG, BMKG, 31 Desember 2020. https://dataonline.bmkg.go.id/home. Accessed 1 Aug 2021
18. Kourentzes, N.: Cran R Project, 16 January 2019. https://cran.r-project.org/web/packages/nnfor/nnfor.pdf. Accessed 31 Aug 2021

Geological Monitoring and Remote Sensing Technology

Development of a Geoinformation Monitoring Module for the Khankalsky Geothermal Deposit in the Chechen Republic

Elina Elsunkaeva(✉), Magomed Mintsaev, and Timur Ezirbaev

Grozny State Oil Technical University named after Acad. M.D. Millionschikov, 100, Isaeva Avenue, Grozny 364051, Russia
elina.elsunkaeva@mail.ru

Abstract. Geothermal energy is one of the renewable energy sources that can become an alternative to increasingly scarce fossil fuels. The Chechen Republic, in particular, the Khankalskoe field has a huge potential of geothermal energy. This research work involves the development, on the basis of a domestic geoinformation systems (further GIS), of a module "Geothermy" (further GIS-module) to monitor the exploited geothermal fields in order to protect the environment from possible negative impacts. The GIS-module contains many functions such as a map view mode, queries, navigation through the monitoring region and displaying the properties of geothermal wells, operational access to databases in order to improve the efficiency of management decision-making. The proposed method in conjunction with geoinformation monitoring of the geological environment, using modern methods and technologies makes it possible to identify zones of dynamic activity of the Earth's surface, areas of thermal water spill, as well as processes associated with the release of carbon dioxide, which will subsequently help prevent the emergence of emergency situations.

Keywords: Geoinformation technology · GIS-module · Database · Integrated monitoring · Geothermal resources

1 Introduction

In recent years, geoinforation monitoring of the geological environment is widely developed, which provides increased informativeness of the conducted set of works, increasing the speed, representativeness and reliability of the information obtained with an overall reduction in its cost. Geoinformation monitoring requires multifactor and multicomponent analysis of the data obtained to assess and solve problem-oriented problems [1, 2].

The territory of the Chechen Republic is one of the most promising geothermal areas in Russia. Having large resources of geothermal water, it is a good base for creating a geoinformation system module (further GIS-module). The GIS-module called by the developers - "Geothermy" will be implemented for geoinformation monitoring of the geological environment, based on modern methods and technologies: geoinformation technologies, ground and remote sensing, laboratory and expeditionary research, etc. [3, 4].

S. Bourennane and P. Kubicek (Eds.): ICGDA 2022, LNDECT 143, pp. 69–77, 2022.
https://doi.org/10.1007/978-3-031-08017-3_6

The importance of the study lies in the fact that the results obtained can be used by the developers of geoinformation monitoring systems for regulatory and supervisory authorities to ensure monitoring of geothermal water deposits, as well as oil and gas fields, deposits of solid minerals, as well as allow to monitor changes in the environmental conditions as a result of man-made impact as a result of development and operation of the above deposits.

The purpose of the study - the development and implementation of methods for maintaining and forming the structure of the current database of indicators of geoinformation monitoring of the near-surface zone of the geological environment on the example of the Khankala field of the Chechen Republic. To achieve this goal it is planned:

1. To analyze the results of research and experience of using geoinformation technology in the development of mineral deposits;
2. Study and characterize the natural features of the Khankala deposit and the possibilities of structuring indicators of geoinformation monitoring of geological environment of geothermal waters in the Chechen Republic;
3. Systematize source materials of geoinformation monitoring (statistical, cartographic, remote sensing, etc.) to form the structure of an up-to-date database of the Geothermal GIS-module;
4. Develop a methodology and criteria for creating a database for multi-factor and multi-component analysis of topical problems;
5. Develop a methodological basis for the logical and hierarchical structure of the database and form the architecture of the GIS-module "Geothermy";
6. Form the structure of "Geothermy" GIS-module for geoinformation monitoring of the geological environment of the Khankala field.

2 Location and General Description of the Study Area

In 2013 Grozny State Oil Technical University named after Acad. M.D. Millionschikov as part of the consortium "Geothermal Resources" began a pilot project to build a geothermal plant at the Khankalsky geothermal field of the Chechen Republic (Russia). The field is located 10 km southeast of the capital Grozny (Fig. 1) [5]. Thermal groundwaters are contained in Middle Miocene Karagan-Chokrak sediments, represented by sandstones with interlayers and lenses of clays. A total of 22 productive strata, ranging in thickness from a few to 60 m, have been identified. The content of thermal groundwater in the Middle Miocene sediments, in addition to favorable filtration parameters of productive strata, determined a high heat flow, structural-tectonic factor, groundwater movement (water is heated in synclinal sags and then rises to the surface) and the lithology of rocks - Karagan-Chokra deposits are located between Sarmatian and Maikop clays, which provide heat storage [6]. The Khankala thermal groundwater deposit is a multilayer reservoir bounded in the northeast and southwest by two strike-slip faults (Fig. 2).

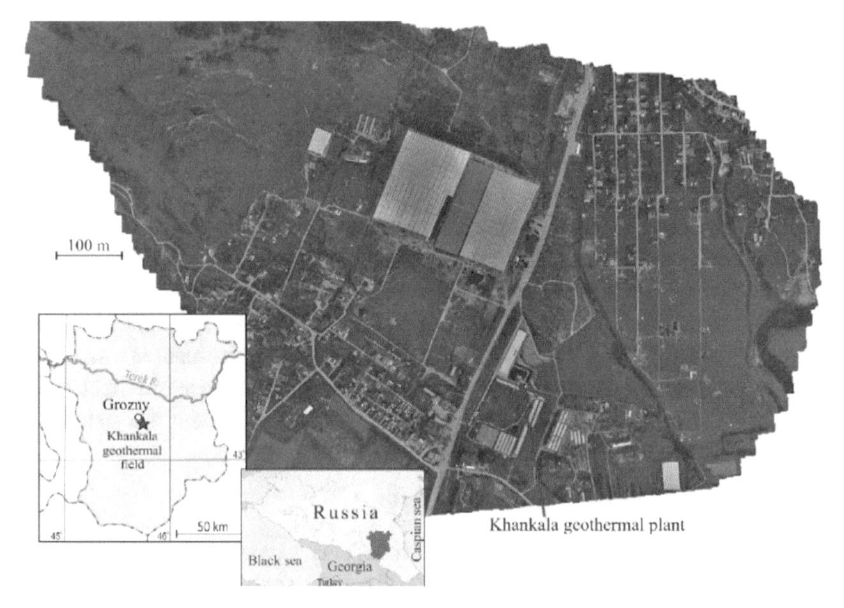

Fig. 1. Khankalskoe thermal groundwater deposit

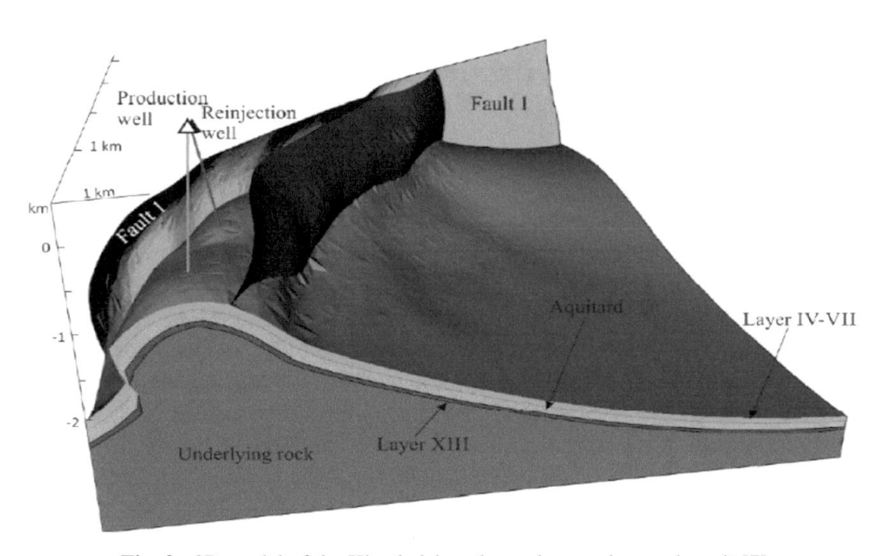

Fig. 2. 3D model of the Khankalskoe thermal groundwater deposit [7]

3 Methods and Methodology

Geoinformation technology is based on electronic maps and thematic databases, which in turn are filled with monitoring data. The use of machine data processing and computer mapping is becoming increasingly widespread, not only in science and technology, but also in management and production activities.

The GIS-module project together with the creation of an actual database for geoinformation monitoring of the geological environment involves the prediction and description of the possible negative impact on the environment during the development and operation of geothermal fields. GIS-module project together with the creation of an actual database for geoinformation monitoring of the geological environment involves the prediction and description of the possible negative impact on the environment from the technogenic impact during the development and operation of geothermal fields.

3.1 Types of Technogenic Impact Monitoring

Monitoring of man-made impacts is divided into several types: a) hydrodynamic monitoring - measurement of levels of formation fluids in control and observation wells, observation of the dynamics of changes in the level of formation fluid in the wellbore of the observation well during pumping during routine geochemical sampling; b) hydrogeochemical monitoring - data from field and laboratory chemical analyses of formation fluid samples from monitoring and observation wells; c) geophysical monitoring - gamma ray (GR), resistivity, neutron porosity, interval transit time and induction electrical log (IEL) resistivity, density (DEN) and litho-density log (LDL), natural gamma ray spectrometry log (SGR), a complex of ground-based geophysical methods.

4 Results

The information of the file system of our module is a composition of rather heterogeneous elements: data of factographic type; geographical maps; information of observation services, literary and other sources; aerospace imagery; normative-reference materials [8]. The "Geothermy" module itself is a structured part of a unified geoinformation system for monitoring the region (Fig. 3). It is intended for database management and editing the main data and directories of the system. The offered module "Geothermy" is intended for formation of a data base, interaction of various levels of system, performance of functions of import and export of the monitoring data, and also preparation of the reporting forms on the data characterizing an ecological situation of objects of monitoring [9–12]. Also it provides an opportunity to store in a hierarchical and logical form all the necessary information of the geoinformation monitoring of the geothermal water field of the studied area.

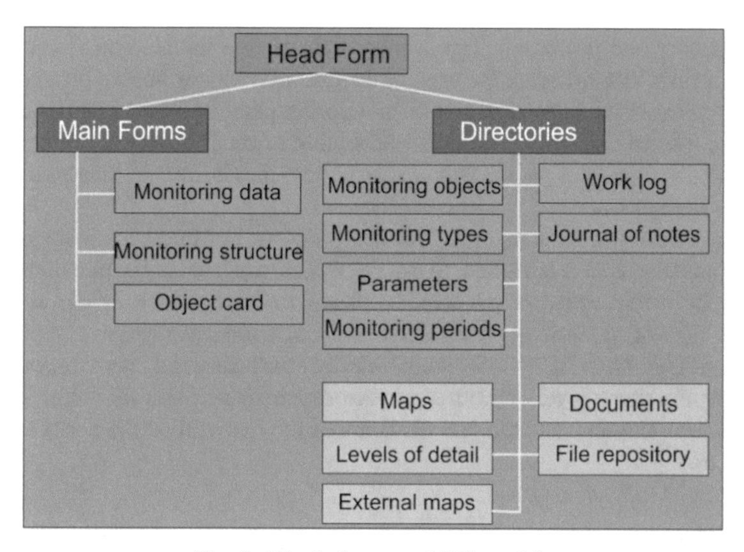

Fig. 3. Block diagram of GIS-module

The logical structure of the module database is some subset of the field monitoring data corresponding to one or more logically connected databases. The logical database has a strictly hierarchical tree structure [13, 14]. The hierarchical structure is represented as a tree consisting of objects of different levels of subordination. The upper level, for example, can occupy one object - a field, the second level - layer, the third - a well, etc. (Fig. 4).

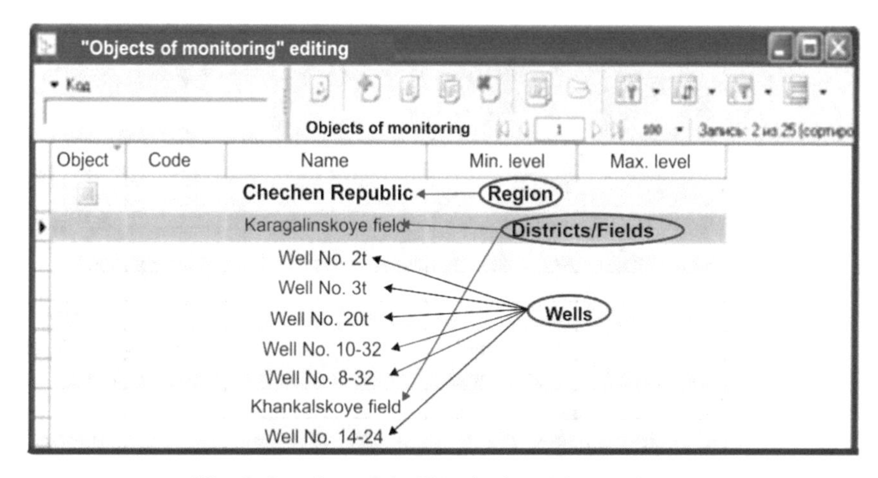

Fig. 4. Interface of the "Monitoring objects" window

To work with the monitoring data, the operations of direct data editing (change of numerical values of parameters) are provided, data import from various sources in the system is performed automatically. The main data of the system is divided into two groups: primary and processed. The primary data are the initial values of monitoring parameters, directly entered by the user, or the data imported from the files received for processing. Processed data are values of monitoring parameters obtained as a result of processing. The tree of system objects, connection of monitoring parameters with the system objects, types and parameters of monitoring, monitoring periods are formed by the user of the system.

The monitoring parameters have a similar property. For them, the belonging to the type of monitoring is also indicated. In this regard, the top level of the tree of monitoring types and parameters are usually located types of monitoring. For example, at the upper level - the type of monitoring “Hydrochemical”, at the second level - the monitoring parameters: “Hydrogen Index”, “Sodium”, etc. For the monitoring periods you can also build a tree, the upper level of the tree of periods are years, the level below - quarters, the third level - months, and the leaf elements of the tree will be the periods of “day” type (Fig. 5).

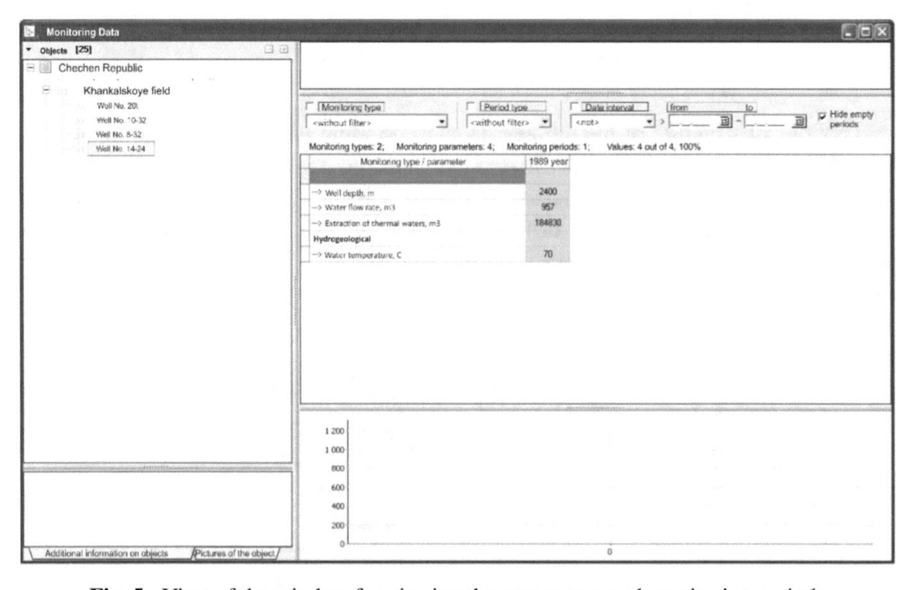

Fig. 5. View of the window for viewing the parameters and monitoring periods

Thermal images obtained with the help of TM and ETM+ sensors by the American Landsat-8 satellite for 2017, 2018 and 2020 have also been added to the system. Temperature readings were calculated using channel 10 Landsat 8.

These thermal satellite images, due to their high resolution and the regularity of surveys, make it possible to trace the dynamics of changes in the temperature field (Fig. 6).

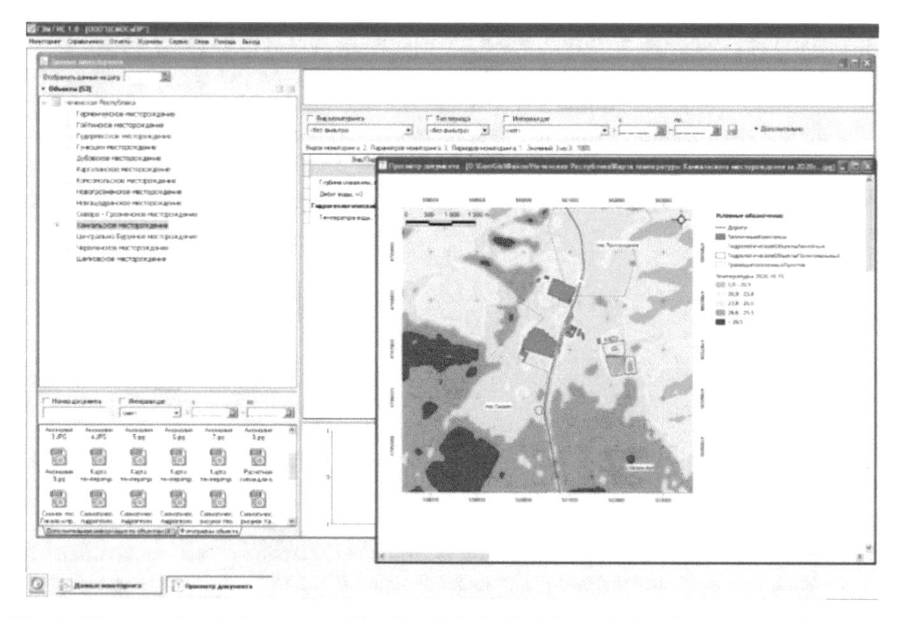

Fig. 6. View window for heat maps of the Khankala field with recalculated temperatures in degrees Celsius with georeferencing to the area loaded into the database

5 Conclusions

During the analytical review of existing GIS-technologies in the development of deposits, the authors concluded that the world and domestic experience of using GIS tools in the development of mineral deposits, suggests that the problem of insufficient number of studies on scientific laws, principles, ideas that constitute the conceptual and organizational and technological basis for managerial decision-making on the operation and development of mineral deposits, is actuality and it should also be noted the insufficient number of developed GIS in Russia directly related to the development and exploitation of geothermal resources, which shows the relevance of this study and determines the task of developing its own GIS module for comprehensive monitoring of the Chechen Republic, its organizational, technological and content information.

In the course of this study, the results of laboratory studies of core samples extracted from wells in the Khankalskoye field, chemical analysis of water obtained from the wells in the study field, as well as geological and geophysical data obtained by researchers in earlier years and finally remote sensing data of the study area in different years (satellite images – 1980–2010s, orthophoto and thermal imaging data - 2018) were systematized.. This allowed to create a problem-oriented structure of the database of the GIS-module "Geothermy" of the geological environment of the Khankalskoye field. The logical and hierarchical structure of the database of the geological environment of the Khankalskoye field in the Chechen Republic was developed and created which ensures quick and qualitative presentation of the monitoring information in the form of tables of storage of the main results with the possibility of its updating in the near-real time scale. The developed methodological framework and criteria for multifactorial and multicomponent analysis of the geological environment of the Khankala field and the proposed system of operational control and response based on geoinformation technology allow to collect, process, analyze, edit and graphically visualize the monitoring data in real time.

In the future, the authors plan to integrate integrated field modelling algorithms into the proposed system field modelling to enable the operator to immediately analyses the obtained monitoring data directly on the ground by displaying a virtual model of the field, the study area.

References

1. Bulayeva, N.M., Gairabekov, I.G., Kerimov, I.A., Mintsayev, M., Ezirbayev, T.B.: Development of geoinformation technologies of complex monitoring of natural resources of the Chechen Republic. Monitor. Sci. Technol. **4**(46), 48–55 (2020)
2. Farkhutdinov, A.M, Cherkasov, S.V, Mintsaev, M.Sh., Shaipov, A.A.: Thermal groundwater of the Chechen Republic: a new stage of use. Nature 25–34 (2017)
3. Farkhutdinov A.M., Cherkasov S.V., Farkhutdinov I.M., Gareev A.M.: Exploitation of the Khankalskoye field of thermal waters. In: Geosphere: Collection of Scientific Articles by Students, Graduate Students and Postgraduate Students of the Department of Geography, vol. 8, pp. 7–8. BashSU, Ufa (2015)
4. Shaipov, A.A., Ezirbaev, T.B., Elsunkaeva, E.V., Ozdieva, T.Kh.: Analysis of remote sensing methodology for geo-environmental monitoring of the Khankala geothermal field. Proceedings of Dagestan State Pedagogical University. Nat. Exact Sci. **12**(4), 103–109 (2018)
5. Elsunkaeva, E.V., Ezirbaev, T.B., Batukaev, A.A., Gadaeva, Z.I.: Monitoring of geothermal resources of the Chechen Republic using geoinformation technology. Monitor. Sci. Technol. **4**(46), 38–47 (2020)
6. Elsunkaeva, E.V., Ezirbaev, T.B., Ozdieva, T.Kh.: Geoecological monitoring of the Khankal field based on thermal infrared imaging. In: IOP Conference Series: Materials Science and Engineering. 3rd International Symposium on Engineering and Earth Sciences (ISEES 2020), p. 012020 (2020)
7. Cherkasov, S.V.: Report on the topic: development of a 3D digital model of a reservoir of geothermal waters. In: Development of an Algorithm for Mathematical modeling of the Temperature distribution within the Reservoir of Geothermal Waters, p. 65 (2013)
8. Ezirbaev, T.B., Elsunkaeva, E.V., Ozdieva, T.Kh.: Monitoring the state of the environment in the areas of influence of industrial enterprises by remote sensing in order to identify the risk of pollution. In: The Second Eurasian RISK-2020 Conference and Symposium. Innovations in Minimization of Natural and Technological Risks. Minimization of the Most Prevalent Project Risks in the Oil and Gas Industry. Abstracts, pp. 116–117 (2020)
9. Kaplunov, Y.V., Limanskiy, A.V., Bulaeva, N.M.: Development of methodological basis for the formation of an actual database of environmental monitoring of the liquidated mines of coal regions of Russia. Monitor. Sci. Technol. **4**, 6–18 (2010)
10. Koshkarev, A.V.: Concepts and terms of geoinformatics and its environment: educational reference manual. In: Russian Academy of Sciences, Institute of Geography, p. 76. Institute of Geography, RAS, Moscow (2000)
11. Lurie, I.K.: Bases of geoinformatics and creation of GIS. In: Berlyant, A.M. (ed.) Remote Sensing and Geographic Information Systems, INEX-92, Part 1, Moscow. p. 140 (2002)
12. Basics of Geoinformatics. In: Tikunov, V.S. (ed.) Textbook. Academia, Moscow (2004)

13. Trifonova, T.A., et al.: Geoinformation Systems and Remote Sensing in Ecological Research, p. 352. Academic Project, Moscow (2005)
14. Farkhutdinov, A.M., Goblet, P., de Foket, C., Cherkasov, S.V.: Computer modeling in the development of thermal energy water reservoirs by the example of the Khankala field. Geothermics **59**, 56–66 (2016)

Mapping Fences in Xilingol Grassland Using High Spatiotemporal Resolution Remote Sensing Data

Tao Liu[1], Xiaolong Liu[2](✉), Libiao Guo[2], and Shupeng Gao[1]

[1] Yunnan Normal University, Kunming 650500, Yunnan, China
[2] Inner Mongolia University of Technology, Hohhot 010051, Inner Mongolia, China
liuxl@mail.bnu.edu.cn

Abstract. As a management tool, fences used in grasslands around the world are becoming increasingly ubiquitous, and the impact on wild species and ecosystems has attracted global attention. However, lacking of large scale accurate fence mapping data has become a limitation of monitoring the influence of fence on ecosystems. In this study, a method is proposed to identify fences based on fused high spatiotemporal resolution remote sensing data in Xilingol grassland, China. The study results indicate that fences as boundaries of adjacent pastures that could be mapped using the fused data by principal component analysis (PCA) feature extraction, multiresolution segmentation and cell edges removal. Tests in our study area for this method showed an overall accuracy of 81.75%, outperformed the single NDVI used result (65.95%).

Keywords: High spatiotemporal resolution · Time series · Image segmentation · Fence

1 Introduction

Fences are now widespread facilities around grazing areas [1], dividing grassland into parcels. Mapping fences is to take a deeper research insight into how fences are affecting the grassland ecosystem. However, mapping fences directly from remote sensing images is still a challenge work. Traditionally, fences are delineated manually using very high resolution (VHR) remote sensing images [2], which is time-consuming, labor-intensive and is not suitable for large spatial extent of fence mapping [3]. Modeling fence location and density method can decrease costs, but having the limitation of general estimate of relative fence density instead of being spatially explicit [4]. In Xilingol grassland, fences are boundaries between parcels, and longtime grazing activities made them apparent boundaries even on medium spatial resolution remote sensing imageries. This means the object-based image segmentation methods could be employed to segment grazing parcels on image into segments to extract grassland boundaries for mapping fences. The object-based image analysis methods are being widely used to identify cultivated field boundary using single VHR remote sensing imageries [5–8]. However, mapping field

S. Bourennane and P. Kubicek (Eds.): ICGDA 2022, LNDECT 143, pp. 78–87, 2022.
https://doi.org/10.1007/978-3-031-08017-3_7

boundaries from such imageries could be challenging, and the acquisition of multitemporal imagery throughout the growing season could overcome the similarity between fields thus is necessary for accurate field boundary delineation [9]. With the impacts of different management (such as grazing routine, mowing or grazing exclusion), vegetation growth status and coverage show discrepancy between pastures of different period within a season in Xilingol grassland (Fig. 1). Therefore, enhancing the discrepancy is the key to extract fences through identifying boundary between pastures.

The objective of this study was to develop a fence identification method based on building NDVI time series by fusing high spatiotemporal remote sensing data and image segmentation, which will produce more accurate fence mapping data.

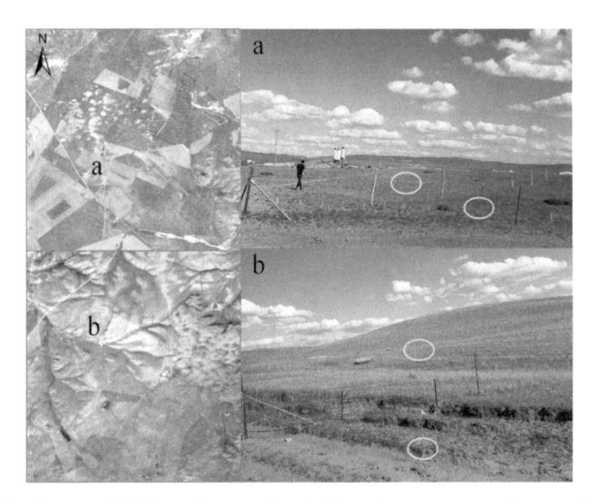

Fig. 1. The fence of Xilingol grassland (the picture was taken on July 2017).

2 Study Area and Data

2.1 Study Area

The study area was located in the Xilingol grassland (43.25 °N–44.25 °N, 115.76 °E–117.12 °E) in the Xilingol League, Inner Mongolia Autonomous Region, China (Fig. 2). The grassland experiences a temperate arid and semi-arid continental climate, with an average annual temperature of 0–3 °C, a freezing period of 5 months, and an average annual precipitation of 200–300 mm. The precipitation is concentrated in summer and autumn. There are high quality natural grasslands suitable for grazing with up to 92% of the total grassland area [10]. Fencing area in Xilingol grassland is about 12.7×10^4 km^2, accounting for 65% of the total grassland area.

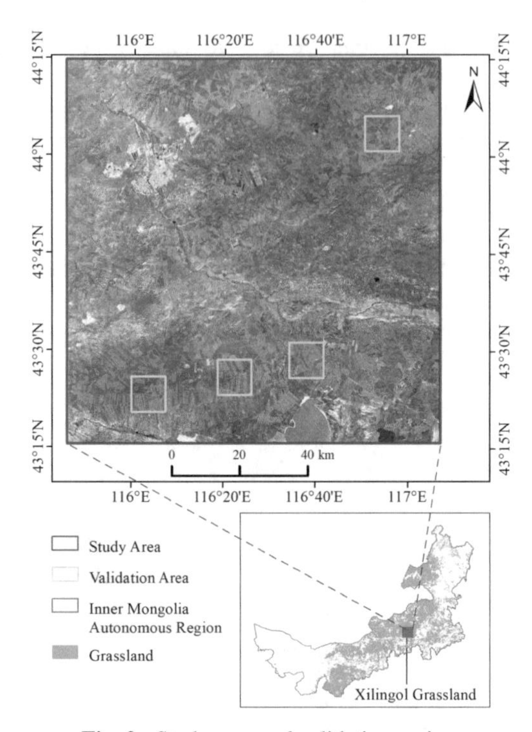

Fig. 2. Study area and validation region.

2.2 Data

Remote Sensing Data. In this study, the Sentinel-2 L1C images with high spatial resolution (10 m) and MODIS images with high temporal resolution (daily) including MOD09GQ (250 m) and MOD09GA (500 m) images were collected. Cloud-free Sentinel-2 L1C data included 5 images of DOY (day of year) 118, DOY 128, DOY 248, DOY 253 and DOY 258 in 2018. MOD09GQ and MOD09GA images were acquired from April to September 2018.

Validation Data. Reference fence (pasture boundary) datasets consisting of four validation areas with 10 $\times$ 10 km was manually digitized by combining ground investigation in July 2017 with very high resolution (50 cm to 30 cm) satellite imagery from the ArcMap base layer. The distribution of the four validation areas is shown in Fig. 2.

3 Methodology

The fences extraction method in this study was based on blended high spatial and temporal remote sensing data and mainly consists of four steps: (1) Sentinel-2 images and MOD09GQ images fusion using the enhanced spatial and temporal adaptive reflectance fusion model (ESTARFM) to build NDVI time series, (2) feature extraction and dimension reduction by PCA to fused high spatiotemporal data, and (3) an image segmentation to the extracted features to achieve primary fence edges by multiresolution algorithm, (4) a cell edges removal to fence extraction results was conducted employing the DP algorithm to smooth the primary fence edges. The detailed algorithm process is shown in Fig. 3.

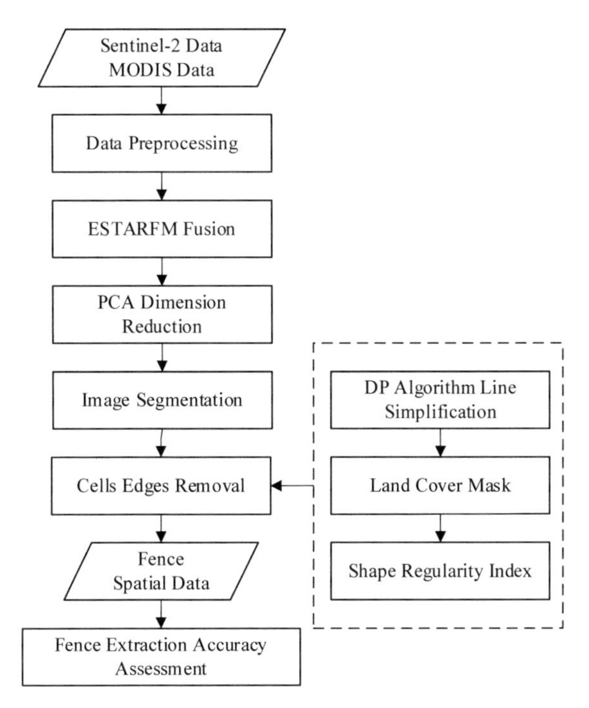

Fig. 3. Flow chart of the fence extraction.

3.1 Spatial and Temporal Data Fusion

MODIS image have high temporal resolution, but low spatial resolution (250 m) which is unsuitable for monitoring heterogeneous landscapes. On the contrary, the Sentinel-2 image has a higher spatial resolution (10 m) which can accurately delimitate the spatial details. However, due to the influence of weather (clouds and fogs), the available observation during the growth season is limited. Considering the landscape complexity of Xilingol grassland and in order to preserve as much spatial and temporal resolution of pastures during growth season as possible, the ESTARFM model, which could accurately predict the surface reflectance and preserve the details in high resolution remote sensing images especially for heterogeneous landscapes [11], was chosen to fuse high spatiotemporal remote sensing data using Sentinel-2 and MOD09GQ images in this study.

3.2 PCA Dimension Reduction and Feature Extraction

The fused high spatiotemporal resolution NDVI data had high dimension. However, due to the similarity of vegetation growing status without significant changes between bands of close dates, there was data redundancy between bands of this data cube (Fig. 4). The PCA has been widely used for temporal analysis of remote sensing data [12]. In this study, PCA was selected to do dimension reduction to preserve the variance and remove redundancy of the NDVI time series data.

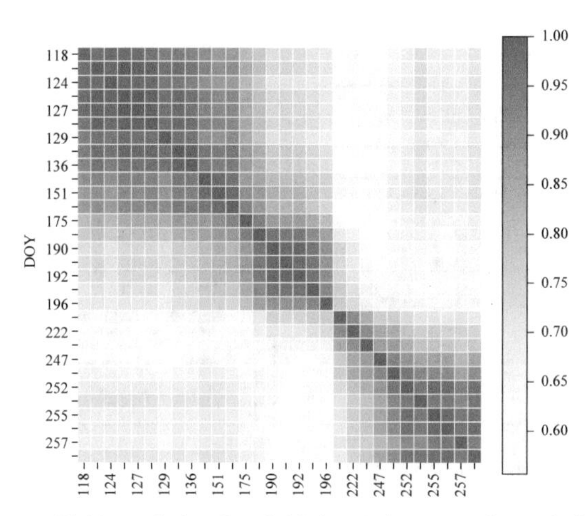

Fig. 4. Correlation coefficient of the fused high spatiotemporal resolution NDVI data by ESTARFM.

3.3 Image Segmentation

In this study, we used the multiresolution segmentation algorithm to generate image segments using the PCA extracted features. The multiresolution segmentation algorithm is a bottom-up method that merges pixels to generate image objects. The algorithm minimizes the average heterogeneity of an image object, but maximizes the difference between nearby objects [13, 14]. To acquire accurate image segmentation results, tests were carried out according to the specific data to get the proper segmentation parameters.

3.4 Cell Edges Removal

The image segmentation results were generated from raster images, which means there will be cell edges in the segmented results and will cause bias to the fence mapping results. In addition, the image segmentation results included both fenced grassland and unfenced land (such as water, desert and impervious land). As shown in Fig. 5.

Fig. 5. Cell edges and unfenced land of image segmentation result.

It is very necessary to do cell edges removal to improve fence mapping, we smooth the image segmentation results and mask unfenced area by the following three steps. (1) To simplify cell edges by removing relatively extraneous vertices while preserving essential shapes, simplification of segmentation result was performed using Douglas-Peucker algorithm (DP) [15, 16] with a 30 m distance threshold. (2) Water, desert and impervious land were masked from DP result with K-Means cluster classification according to phenological variation of NDVI of image objects. (3) We developed a shape regularity index [17] to remove pseudo boundaries. Shape regularity index was the weighted sum of combining maximum notch ratio (N_r), maximum length of a polygon (L_{max}), the increment of perimeter between polygon and its convex hull (D_{prei}), and the increment of area between the polygon and its smallest enclosing circle (D_{area}). The weight metrics determined by PCA. The mathematical formula of shape regularity index is as follows:

$$S = W_1 \times N_r + W_2 \times L_{max} + W_3 \times D_{prei} + W_4 \times D_{area} \tag{1}$$

$$N_r = N_i / N_{max} \tag{2}$$

$$D_{prei} = (P_i - P_h)/P_i \tag{3}$$

$$D_{area} = (A_c - A_i)/A_c \tag{4}$$

where S is the shape regularity index; W_1 to W_4 are weight metrics; N_i is the number of notches for the ith image object, and N_{max} is the maximum value of notches for image objects in our study area; P_i and P_h are perimeters of the ith image object and its convex hull respectively; A_i, A_c are areas of the ith polygon and the smallest enclosing circle respectively.

3.5 Accuracy Assessment

In this study, edge-based metrics are calculated to assess accuracy using length of extracted fence and its corresponding reference fence in our validation areas, which include overall accuracy (OA), commission error (CE) and omission error (OE). A higher omission error rate suggests that there were more fence boundaries missed from the reference fence boundaries. CE indicates false fences extracted, often due to over segmentation.

4 Results and Analysis

4.1 Image Fusion Accuracy

According to the scatterplot of ESTARFM fusion and observed result of DOY 253 in 2018 (see Fig. 6), the determination coefficient (R^2) was greater than 0.7, which indicated that the fused result was in high correlation with the observed result. The NDVI time series was built by using Sentinel-2 images and the ESTARFM fused results (Fig. 7) in this study.

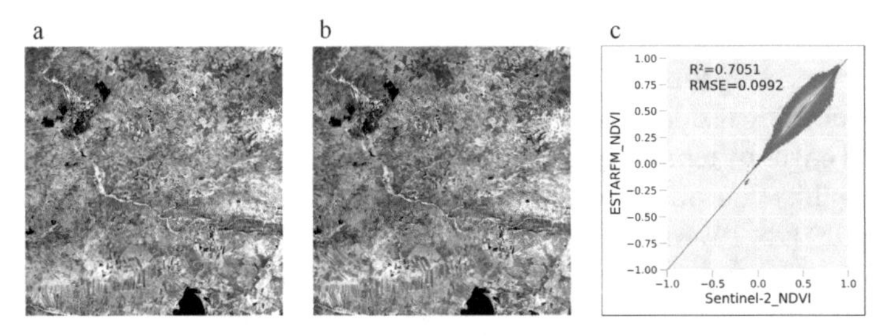

Fig. 6. ESTARFM fusion result of DOY 253 in 2018: a. ESTARFM result; b. Observed Sentinel-2 data; c. Scatterplot of the ESTARFM result and the observed Sentinel-2 data.

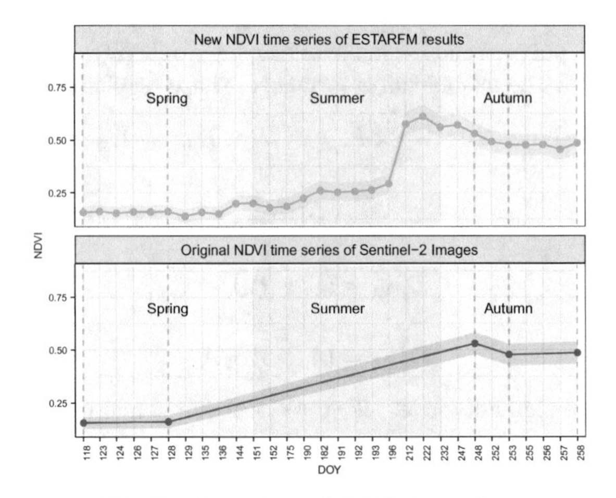

Fig. 7. Comparison of NDVI time series.

4.2 Fence Extraction Result and Accuracy Assessment

We did a comparison between two fence extraction schemes, the single NDVI strategy and the time series strategy. Image of DOY 258 with a high discrimination between pastures was selected as reference to evaluate the fence extraction performance of the method given in this study. Using multiresolution segmentation, DP line simplification, land cover mask and shape regularity index, we obtained the fence extraction results of DOY 258 in 2018. The accuracy metrics obtained over the four validation areas using time series NDVI and single NDVI are shown in Table 1, where time series NDVI result shows a better performance with a mean OA 83.75%, OE 18.83 and CE 17.67 from all validation areas. It is apparent that time series NDVI can achieve better fence extraction results than single NDVI. The fence mapping results are as shown in Fig. 8.

Table 1. Fence extraction accuracy results for the four validation areas of NDVI time series and single NDVI image.

Validation area	NDVI time series			Single NDVI image		
	OA (%)	OE (%)	CE (%)	OA (%)	OE (%)	CE (%)
S1	80.48	22.41	16.62	65.95	43.91	24.20
S2	82.09	16.41	19.42	74.24	30.57	20.95
S3	84.13	15.94	15.80	72.30	34.47	20.94
S4	80.30	20.58	18.83	75.80	29.24	19.17
Mean	81.75	18.83	17.67	72.07	34.55	21.31

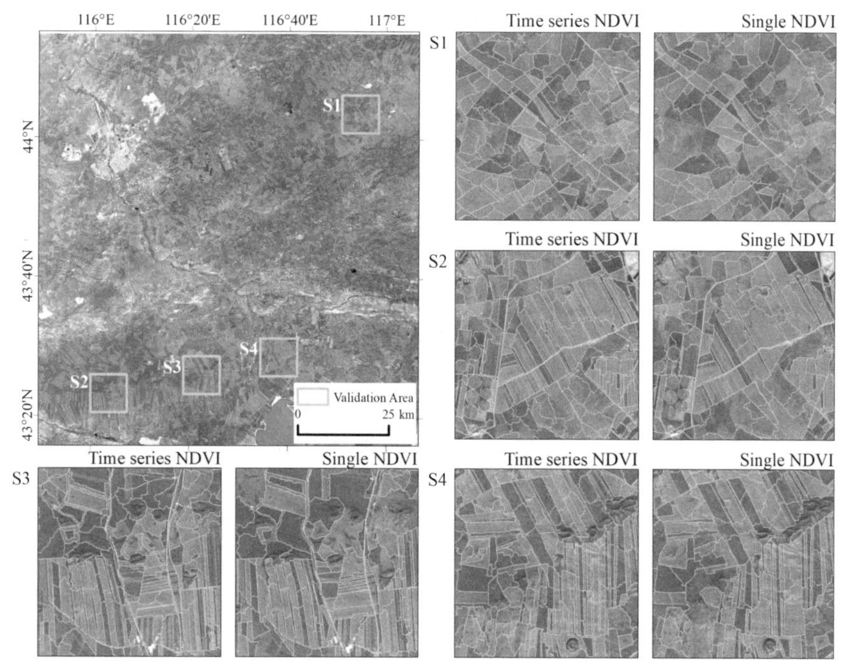

Fig. 8. Results of fence extraction for the four validation areas of time series NDVI and single NDVI.

5 Conclusion

In this study, a fence mapping based on fused high spatiotemporal resolution remote sensing data method was proposed. The performances of the proposed method were evaluated in Xilingol grassland of Inner Mongolia Autonomous Region, China. The fused high spatiotemporal resolution NDVI outperformed the single NDVI data used result with an overall accuracy of 81.75%, which indicates the proposed method could achieve a good performance for fence extraction.

Our research shows that time series data can enhance the spectral difference between adjacent grassland parcels. Although we have collected all available cloud-free Sentinel-2 images during the growing season of 2018 and used the fusion model to obtain more images, the large number of unavailable images caused by cloud or fog coverage is still the main obstacle in building a complete optical remote sensing time series during this period. The tunning of the multiresolution segmentation parameter will cause uncertainties to the result of the proposed method.

In further research, we will collect more usable images to construct the complete time series, and will try to modify the segmentation method. Furthermore, the proposed method will apply in other study areas to improve the robustness.

References

1. Woodroffe, R., Hedges, S., Durant, S.M.: To fence or not to fence. Science **344**(6179), 46–48 (2014)
2. Jakes, A.F., Jones, P.F., Paige, L.C., et al.: A fence runs through it: a call for greater attention to the influence of fences on wildlife and ecosystems. Biol. Conserv. **227**, 310–318 (2018)
3. Mueller, M., Segl, K., Kaufmann, H.: Edge-and region-based segmentation technique for the extraction of large, man-made objects in high-resolution satellite imagery. Pattern Recogn. **37**(8), 1619–1628 (2004)
4. McInturff, A., Xu, W., Wilkinson, C.E., et al.: Fence ecology: frameworks for understanding the ecological effects of fences. Bioscience **70**(11), 971–985 (2020)
5. Li, D., Zhang, G., Wu, Z., Yi, L.: An edge embedded marker-based watershed algorithm for high spatial resolution remote sensing image segmentation. IEEE Trans. Image Process. **19**(10), 2781–2787 (2010)
6. Witharana, C., Civco, D.L.: Optimizing multi-resolution segmentation scale using empirical methods: exploring the sensitivity of the supervised discrepancy measure Euclidean distance 2 (ED2). ISPRS J. Photogramm. Remote. Sens. **87**, 108–121 (2014)
7. Kavzoglu, T., Tonbul, H.: A comparative study of segmentation quality for multi-resolution segmentation and watershed transform. In: 8th International Conference on Recent Advances in Space Technologies (RAST), Istanbul, Turkey, pp. 113–117. IEEE (2017)
8. Munyati, C.: Optimising multiresolution segmentation: delineating savannah vegetation boundaries in the Kruger National Park, South Africa, using Sentinel 2 MSI imagery. Int. J. Remote Sens. **39**(18), 5997–6019 (2018)
9. Watkins, B., Van Niekerk, A.: Automating field boundary delineation with multi-temporal Sentinel-2 imagery. Comput. Electron. Agric. **167**, 105078 (2019)
10. Chi, D., Wang, H., Li, X., et al.: Assessing the effects of grazing on variations of vegetation NPP in the Xilingol Grassland, China, using a grazing pressure index. Ecol. Indic. **88**, 372–383 (2018)

11. Zhu, X., Chen, J., Gao, F., et al.: An enhanced spatial and temporal adaptive reflectance fusion model for complex heterogeneous regions. Remote Sens. Environ. **114**(11), 2610–2623 (2010)
12. Adami, M., Bernardes, S., Arai, E., et al.: Seasonality of vegetation types of South America depicted by moderate resolution imaging spectroradiometer (MODIS) time series. Int. J. Appl. Earth Observ. Geoinf. **69**, 148–163 (2018)
13. Cheng, G., Han, J.: A survey on object detection in optical remote sensing images. ISPRS J. Photogramm. Remote Sens. **117**, 11–28 (2016)
14. Tetteh, G.O., Gocht, A., Conrad, C.: Optimal parameters for delineating agricultural parcels from satellite images based on supervised Bayesian optimization. Comput. Electron. Agric. **178**, 105696 (2020)
15. Douglas, D.H., Peucker, T.K.: Algorithms for the reduction of the number of points required to represent a digitized line or its caricature. Cartographica Int. J. Geogr. Inf. Geovis. **10**(2), 112–122 (1973)
16. Liu, X., Shi, Z., Huang, G., et al.: Time series remote sensing data-based identification of the dominant factor for inland lake surface area change: anthropogenic activities or natural events? Remote Sens. **12**(4), 612 (2020)
17. Yan, H.: Quantitative relations between spatial similarity degree and map scale change of individual linear objects in multi-scale map spaces. Geocarto Int. **30**(4), 472–482 (2015)

Comparative Study on Remote Sensing Image Classifier of Jiulong River Basin

Yiping Liao[1], Guangsheng Liu[1](✉), Haijun Luan[2,3], Meiduan Zheng[1], and Guojiang Deng[1]

[1] Fujian Engineering and Research Center of Rural Sewage Treatment and Water Safety, Xiamen University of Technology, Xiamen 361024, China
liugs@xmut.edu.cn

[2] College of Computer and Information Engineering, Xiamen University of Technology, Xiamen 361024, China

[3] Institute of Big Data for Natural Hazards Monitoring of Fujian Development and Reform Commission, Xiamen University of Technology, Xiamen 361024, China

Abstract. Remote sensing technology has gradually become the mainstream means of dynamic monitoring of the large-scale surface environment by virtue of its large-scale, low-cost, and high-efficiency advantages. Based on the Google Earth Engine (GEE), using Landsat and Sentinel satellite images from 1990 to 2018 as the data source, selecting the Jiulong River Basin as the research area, calling CART (Classification And Regression Tree), Random Forest, Naive Bayes, Minimum Distance, Support Vector Machine and carry out accuracy evaluation and result analysis of the classification results. The results show that the land cover classification can be completed quickly based on the GEE platform, and each classification method can achieve better results; the support vector machine algorithm extraction results are the best, the overall accuracy is above 78%, and the Kappa is above 0.75; vegetation and field land is the main land type of Jiulong River Basin.

Keywords: Remote sensing · Precision assessment · Classifier · Land use change

1 Introduction

Remote sensing has gradually become the mainstream means of large-scale surface environmental dynamic monitoring by virtue of its advantages of large-scale, low-cost, and wide coverage area [1–3]. The newly developed geospatial data analysis cloud platform GEE (Google Earth Engine) has changed the traditional remote sensing processing method. Its huge remote sensing image data set and high-performance computing capabilities have solved the problem of remote sensing image acquisition [4, 5]. The GEE cloud platform directly provides a variety of image interpretation methods including CART (Classification and Regression Tree), Random Forest, Naive Bayes, Minimum Distance, Support Vector Machine, etc. [6]. Limited by the climate of the study area and the attributes of satellites, selecting the most suitable classification method is a major difficulty in remote sensing of the surface extraction [7, 8].

S. Bourennane and P. Kubicek (Eds.): ICGDA 2022, LNDECT 143, pp. 88–94, 2022.
https://doi.org/10.1007/978-3-031-08017-3_8

Jiulong River is located in the southern part of Fujian Province and is the second largest river in Fujian Province. It has a drainage area of 14,741 km^2, accounting for about 12% of the land area of Fujian Province. At present, most studies on the Jiulong River Basin are limited to the confluence of the Jiulong River Basin, and there are few observations on the surface environmental factors of the Jiulong River Basin on a large scale and long time series [9, 10].

This research takes the Jiulong River basin as the research area, based on the Landsat (5, 8) and Sentinel-2 series of images provided by the GEE cloud platform and the CART, Random Forest, Naive Bayes, Minimum Distance, Support Vector Machine, etc. are used to classify the land cover of the remote sensing images of the Jiulong River Basin from 1990 to 2018, and to explore the land classification method that is most compatible with the Jiulong River Basin, with a view to the surface environment of the Jiulong River Basin Long-term dynamic monitoring contributes.

2 Methods

2.1 Study Data

Using Landsat remote sensing images and Sentinel-2 remote sensing images as the main data sources, the image data comes from the online data of the GEE cloud platform, and select the 7 periods of the issue from 1990 to 2018 (1990, 1995, 2000, 2005, 2008, 2014, 2018) As a case study, the images used include Landsat 5 TM, Landsat 8 OIL, and Sentinel-2. In order to weaken the impact of seasonal changes on land use changes, remote sensing image maps in the first half of the year were selected.

2.2 Classification

Samples are selected by analyzing the features of the ground features in the study area and visual interpretation methods. Among them, 70% of the samples are used for land classification, and 30% of the samples are used as the samples required for accuracy testing. A total of 7 phases of Landsat and Sentinel images in the Jiulong River Basin from 1990 to 2018 are classified to extract land cover information. In order to ensure the accuracy of the research results, a total of 5 current mainstream land classification algorithms were selected for this study. The following are the basic principles of the 5 algorithms.

CART. CART assumes that the decision tree is a binary tree, and the values of the internal node features are "yes" and "no", the left branch is the branch with the value "yes", and the right branch is the branch with the value "no". Such a decision tree is equivalent to recursively dicing each feature, dividing the input space, that is, the feature space into a finite number of units, and determining the predicted probability distribution on these units.

Random Forest. Random forest is a classifier that contains multiple decision trees, and its output category is determined by the mode of the output category of individual trees. The randomness of random forest is reflected in that the training samples of each tree are random, and the split attribute set of each node in the tree is also randomly selected and determined. With these two random guarantees, random forests will not produce overfitting.

Naive Bayes. The Bayesian principle is based on the Bayesian principle and uses knowledge of probability statistics to classify sample data sets. The naive Bayes method is based on the Bayes algorithm, which is simplified, that is, it is assumed that the attributes are conditionally independent of each other when the target value is given. Although this simplification method reduces the classification effect of the Bayesian classification algorithm to a certain extent, in actual application scenarios, it greatly simplifies the complexity of the Bayesian method.

Minimum Distance. The minimum distance classification is the most basic classification method in the classifier. It calculates the distance D from the unknown category vector X to the center vector of each category (such as A, B, C, etc.) known in advance, and then the vector X to be classified is attributed to the classification method of the smallest one of these distances.

Support Vector Machines. Support vector machine is proposed for the binary classification problem, and successfully applied the sub-solution function regression and the first-class classification problem. Its classification algorithm uses linear calculations to treat samples of a certain category as one category, and the remaining samples of other categories as another category. This becomes a two-class classification problem. Repeat the above steps in the remaining samples.

2.3 Precision Assessment

Confusion Matrix. Confusion matrix is a standard form of error expression, reflecting the expression of the classification results to the actual ground category. It can not only represent the total error of each category, but also the error of the category. The error matrix is generally composed of n × n matrix columns to indicate the accuracy of the classification result, where n represents the number of categories, and the number on the diagonal represents the data in the correct classification of the sample, the numbers in other positions are the numbers in the wrong classification of the sample.

Overall Accuracy. The overall accuracy is the ratio of the total number of correct samples in the test sample to the total number of all samples, and the formula is shown in 1.

$$OA = \frac{\sum_{i=1}^{n} P_{ii}}{P} \tag{1}$$

where P_{ii} is the number of samples in the row i and column i of the confusion matrix (that is, the number of samples that are correctly classified) and P is the total number of samples used for accuracy evaluation.

Kappa. When the overall accuracy index is used to evaluate the classification accuracy, its objectivity depends on the sampling sample and the method, and the classification result cannot be evaluated well. The Kappa coefficient will use most of the data in the confusion matrix, which can fully reflect the overall accuracy of the image classification. The formula is shown in 2. Its value is between -1 and 1. The closer to 1, the more accurate the classification.

$$K = \frac{P\sum_{i=1}^{n} P_{ii} - \sum_{i=1}^{n}(P_{i+} \cdot P_{+i})}{P^2 - \sum_{i=1}^{n}(P_{i+} \cdot P_{+i})} \tag{2}$$

where n is the total number of columns in the confusion matrix, the number of categories; P_{ii} is the number of samples in the i row and column i of the confusion matrix; P_{i+} and P_{+i} are the total number of pixels in the row i and column i, respectively; P is the total number of pixels used for accuracy evaluation.

3 Results and Analysis

3.1 Precision Analysis of Image Classification in Jiulong River Basin

Remote sensing images of Jiulong River Basin in 1990, 1995, 2000, 2005, 2008, 2014 and 2018 were selected for processing in this study, among which Landsat 8 OIL series images in 2014 and Sentinel-2A series images in 2018 were selected. The rest were derived from Landsat 5 TM series images. The same time and accuracy were evaluated using the five classification methods described in Sect. 2.2. Table 1 reveals the overall accuracy of images of 7 years calculated by using 5 classification methods. As shown in Table 1, the sentinel-2 remote sensing images in 2018 were classified by CART, random forest method and support vector machine. The overall accuracy of classification results was high, and the gap between them was small. Naive Bayes and Minimum Distance classification have poor adaptability to Sentinel-2 satellite images, and their accuracy is lower than the other three classification methods. For Landsat series of satellites, the overall accuracy of Support Vector Machine classification is above 78%, which is higher than the other four classification methods. Based on the analysis of the overall accuracy of classification results, Support Vector Machine classification of Sentinel-2 and Landsat images in Jiulong River Basin are the best.

Table 2 reveals the Kappa calculated by confusion matrix after using five classification methods during 1990–2018, which can more comprehensively reflect the overall accuracy of classification. As shown in Table 2, Sentinel-2 satellite uses Support Vector Machine, and its classification result Kappa is higher than the other four classification methods. For Landsat series satellites, the Kappa of Support Vector Machine classification are all above 0.75, which means that the classification results of Landsat images processed by Support Vector Machine classification in this study have achieved quite satisfactory accuracy, and are always higher than the other four classification methods. In general, according to Kappa coefficient analysis, if Sentinel-2 and Landsat satellite images are processed simultaneously, Support Vector Machine classification has the best effect.

Table 1. Overall accuracy of classifier from 1990 to 2018

	CART	Random Forest	Naive Bayes	Minimum distance	Support vector machine
1990	64.92%	74.81%	73.41%	69.44%	78.89%
1995	76.47%	76.22%	76.98%	72.68%	81.82%
2000	61.57%	85.82%	74.79%	68.45%	86.47%
2005	69.11%	81.02%	75.13%	68.74%	80.54
2008	60.10%	78.01%	74.96%	67.97%	83.47%
2014	71.93%	65.51%	70.23%	54.16%	82.40%
2018	89.96%	87.09%	64.61%	69.79%	90.54%

Table 2. Kappa coefficients of classifier from 1990 to 2018

	CART	Random Forest	Naive Bayes	Minimum distance	Support vector machine
1990	0.63	0.69	0.59	0.55	0.77
1995	0.73	0.73	0.63	0.57	0.75
2000	0.54	0.83	0.70	0.62	0.84
2005	0.64	0.76	0.63	0.62	0.74
2008	0.54	0.69	0.72	0.61	0.81
2014	0.69	0.54	0.60	0.48	0.78
2018	0.79	0.74	0.49	0.59	0.89

3.2 Support Vector Machine Classification Result

Based on high-resolution remote sensing images to analyze the actual land cover situation of Jiulong River Basin, combined with the statistical yearbook of Zhangzhou city and Longyan City and other reference materials. In addition to water body, vegetation and construction land, agriculture and fishery are the main primary industries in Jiulong River Basin, and unused land can effectively reflect the progress of regional urbanization. Therefore, the land cover classification system of this study is determined, which mainly includes water body, vegetation, construction land, cultivated land, aquaculture land and unused land. Support Vector Machine method is used for classification.

Figure 1 shows the spatial distribution of land categories in Jiulong River Basin in 1990, 1995, 2000, 2005, 2008, 2014 and 2018 by support vector machine classification method and reclassified in Arcgis 10.6. As shown in Fig. 1, land use distribution characteristics of Jiulong River Basin over the years can be intuitively observed. This shows that remote sensing technology can well process large scale and long time series images of the study area. The classification results of seven years indicate that vegetation and cultivated land are the main land in Jiulong River Basin. Both are distributed in various forms in various locations in the Jiulong River basin. Construction land and unused land are increasing year by year, and the distribution of the two land types is relatively concentrated and close to the distribution of water body. The main changes are construction land and arable land. The change of construction land area is the largest, showing a steady increasing trend, and its distribution is relatively concentrated and close to water. The change of cultivated land area showed a decreasing trend, but the decreasing amplitude was small, and the spatial change was mainly manifested by the gradual fragmentation of classified images. In general, the pattern of farmland dominant in Jiulong River Basin has not changed, and it is predicted that the pattern will not change too much in the future.

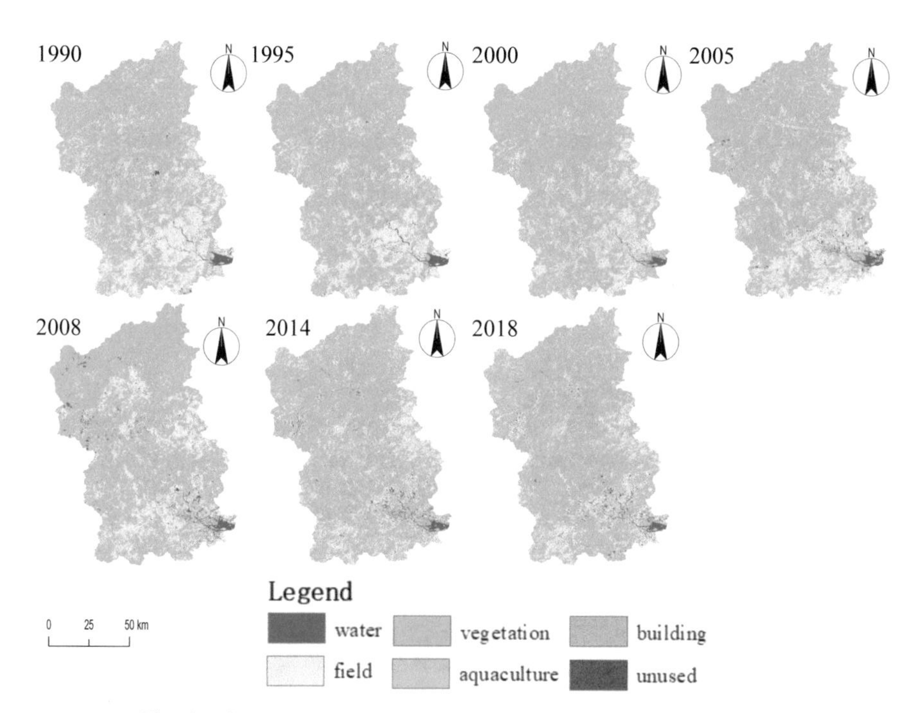

Fig. 1. 1990–2018 Land spatial characteristics in Jiulong River Basin

4 Conclusions

Based on the GEE cloud platform, the Landsat satellite and Sentinel satellite images were acquired, and the various classification methods provided by it were used to process the remote sensing images of the Jiulong River Basin on the two satellites from 1990 to 2018, and the following conclusions were obtained:

(1) Among the five classification methods, when processing Sentinel-2 satellite images of the Jiulong River Basin, the overall classification accuracy and Kappa coefficient of support vector machine classification are always higher than the other four classification methods. When processing Landsat 5 TM satellites or Landsat 8 OIL

from 1990 to 2014, the overall accuracy of support vector machine classification and Kappa are relatively stable, so that the classification results maintain quite satisfactory accuracy. The classification accuracy of Random Forest classification is second only to SVM classification, but for images of different years the classification accuracy is not stable. In general, Sentinel-2 satellite and Landsat satellite images of the Jiulong River Basin are processed with Support Vector Machines for the best classification.

(2) In the past 30 years, vegetation and arable land in the Jiulong River Basin have always been the main land types, covering the entire watershed. Construction land and unused land show a trend of increasing year by year. However, the area of cultivated land showed a decreasing trend, with a small decrease.

Acknowledgements. This work was funded by National natural science foundation of China (No. 51809222) and the Natural Science Foundation of Fujian Province, China (No. 2020J01261) and the "Scientific Research Climbing Plan" Project from Xiamen University of Technology (No. XPDKT19010).

References

1. Xu, D., et al.: J. Clean. Prod. **321**, 128948 (2020)
2. Rawat, J.S., Biswas, V., Kumar, M.: Egypt. J. Remote Sens. Space Sci. **16**(1), 111–117 (2013)
3. Xie, Z., Tang, L., Huang, Y., Huang, J.: J. Subtrop. Resour. Environ. **16**(02), 1–9 (2016)
4. Gorelick, N., Hancher, M., Dixon, M.: Remote Sens. Environ. **202**, 18–27 (2017)
5. Liu, Y.Y., Tian, T., Zeng, P.: Chin. J. Appl. Ecol. **31**(9), 3163–3172 (2020)
6. Zou, D., Li, X., Kang, R., Luo, J.: Geomat. Spat. Inf. Technol. **44**(S1), 100–102 (2021)
7. Su, Z.: Land Dev. Eng. Res. **5**(01), 1–5 (2020)
8. Liang, M.: Beijing Surv. Mapp. **32**(12), 1512–1516 (2018)
9. Rawat, J.S., Kumar, M.: Egypt. J. Remote Sens. Space Sci. **18**(1), 77–84 (2015)
10. Huang, H., Lin, C., Yu, R., Yan, Y., Hu, G., Li, H.: RSC Adv. **9**(26), 14736–14744 (2019)

OSM-GAN: Using Generative Adversarial Networks for Detecting Change in High-Resolution Spatial Images

Lasith Niroshan(✉) and James D. Carswell

Technological University Dublin, Dublin, Ireland
d19126805@mytudublin.ie, james.carswell@tudublin.ie

Abstract. Detecting changes to built environment objects such as buildings/roads/etc. in aerial/satellite (spatial) imagery is necessary to keep online maps and various value-added LBS applications up-to-date. However, recognising such changes automatically is not a trivial task, and there are many different approaches to this problem in the literature. This paper proposes an automated end-to-end workflow to address this problem by combining OpenStreetMap (OSM) vectors of building footprints with a machine learning Generative Adversarial Network (GAN) model - where two neural networks compete to become more accurate at predicting changes to building objects in spatial imagery. Notably, our proposed OSM-GAN architecture achieved over 88% accuracy predicting/detecting building object changes in high-resolution spatial imagery of Dublin city centre.

Keywords: Change detection · Remote sensing · OpenStreetMap · Generative Adversarial Networks · GIS

1 Introduction

Geospatial change detection functionality is an acknowledged component of many practical GIS applications, for example, urban planning, natural disaster prediction, agricultural monitoring, etc. As such, various approaches have been explored and employed to automatically recognise changes in spatial imagery over time. These range from basic statistical implementations to traditional image processing techniques to more complex Deep Learning approaches. In urban cases, successfully obtaining accurate change detection results depends highly on imagery resolution, as lower resolution images can obfuscate a significant amount of important ground object detail – e.g., the precise edges and intersections of buildings in a crowded urban setting.

However, detecting feature/object changes in aerial/satellite (spatial) imagery is a challenging task due to many factors – e.g., a general lack of easily attainable/freely available high-resolution spatial imagery, automating the complex object (e.g., building) extraction process, and the comparatively modest accuracy (±80%) of current change detection algorithms applied to this domain. In support of this study, a customised spatial image *crawler* was developed to search for freely available *Google Earth* and *Bing Maps*

S. Bourennane and P. Kubicek (Eds.): ICGDA 2022, LNDECT 143, pp. 95–105, 2022.
https://doi.org/10.1007/978-3-031-08017-3_9

satellite images from various sources at different spatial resolutions. Once obtained, this raster data is merged with *OpenStreetMap* (OSM) vectors to train the novel OSM-GAN change detection mechanism described in this paper.

At present, Convolutional Neural Network (CNN) models are used a great deal in Artificial Intelligence (AI) applications for resolving general image processing and classification tasks. Among the various *Deep Learning* (a subfield of Machine Learning) methods, Generative Adversarial Networks (GAN) have recently been developed to learn (train) a function (model) that maps (translates) an *input* image to an *output* image. Over the past five years, the task of translating one possible representation of data into another, such as image-to-image translation, has become a common application for GANs. As an example, Isola et al. proposed a general-purpose adversarial network solution in 2017 for image-to-image translation named *Pix2Pix* [1].

Our approach applies the *Pix2Pix* image translation technique to predict/identify possible changes to building objects in high-resolution (30cm/pixel) satellite images. To begin, we must first convert OSM building footprint data (vector) to raster format for use as an output image - since *Pix2Pix* image translation expects both input and output images in raster format for training purposes. Figure 1 shows a real-world (Dublin) example of one *Pix2Pix* training set used to train our OSM-GAN model.

Fig. 1. A joined Pix2Pix raster training sample used for learning the OSM-GAN model. The left side is the input Google satellite image – the right side is the current OSM building footprint output image (feature-map) of the same Dublin area.

This paper describes how online crowdsourced spatial data can be utilised successfully in state-of-the-art Machine Learning applications. In it, we propose an automated change detection framework for OSM buildings that exploits GAN image-to-image translation techniques. The paper is organised as follows: Sect. 2 covers some background and related work on this topic; Sect. 3 explains our proposed OSM-GAN methodology in some detail; Sect. 4 reports on experimental results followed by some Conclusions with a brief discussion on plans for future work.

2 Background and Related Work

This section reviews some related background work relevant to our approach, including an overview of GANs, Conditional GANs, and other noteworthy change detection mechanisms in the literature. For example, applications and improvements to the GAN methodology have increased significantly over time, with different types of GAN frameworks developed for various purposes:

- CapsGAN to generate 3D images with various geometric transformations [3]
- GANSynth to produce audio streams [4]
- GauGAN to transform doodles into highly realistic landscapes [5]
- StyleGAN to generate more realistic images (e.g., human faces, cars, and rooms) [6]
- ChemGAN for drug discovery [7]

2.1 Generative Adversarial Networks

Generative Adversarial Networks (GAN) were proposed by Goodfellow et al. in 2014 as a new class of Machine Learning models where two separate models compete against each other as if in a game, e.g., chess/backgammon/etc. [2]. The basic GAN works like a *minimax* recursive algorithm to find the optimal move for a player. One model is called the *Generator* (G), and the other is called the *Discriminator* (D). Briefly, the Generator trains a generative model to generate fake data similar to the real *feature-map* (right side of the training set image) from a random noise vector (array of 0 s and 1 s) as input. Conversely, the adversary Discriminator is trained to classify/distinguish between the generated (fake) feature-map and the ground truth (real) feature-map.

2.2 Conditional Generative Adversarial Networks

Mirza and Osindero introduced *Conditional Generative Adversarial Networks* (CGAN) also in 2014 [8]. The significant improvement of CGAN over GAN is the addition of a conditional state to the output generation, as usually there is no control over generating output in a GAN. CGAN includes a *condition* (uses both left and right sides of the training sample simultaneously) as input to the Generator and Discriminator to help resolve the issue of an image being real or fake. As it happens, including a condition (feature-map) in the training sample input to the Discriminator function results in a more accurate method for identifying real images – if there is a building in the satellite image, there should also be a predicted building in the resulting feature-map.

Image-to-Image Translation. The main idea behind *image-to-image translation* is that a given input image (e.g., a sketch/outline of an object) translates or transforms into another higher-level representation (e.g., a photo-realistic image) of the set of input information. Therefore, many computer vision and image processing problems (e.g., edge detection, object localisation, sketch-to-photo translation, etc.) can be interpreted as a form of image-to-image translation.

Isola et al. [1] presented several generalised uses of Conditional GAN based image-to-image translation such as labels-to-street scenes, black & white images-to-colour images, sketches-to-photos, style transfer applications, and especially aerial images-to-maps, the main focus of this study. *Pix2Pix* is their implementation of image-to-image translation, which is freely available for use in *GitHub.*[1] We use an updated version of *Pix2Pix* in the OSM-GAN experiments carried out in this work.

2.3 Detecting Spatial Changes in Spatial Images

An accurate change detection mechanism can initiate many other advanced geo-analytic research applications in the GIS domain – where the challenge of change detection has been investigated many ways over the years to address various mapping problems. For example, a considerable number of image processing and computer vision approaches have been introduced for temporal change detection in spatial images, such as Markov random fields [9], Principal Component Analysis [10], CNN based difference image approach [11], and Recurrent neural network-based U-Net models [12].

Recently (2018), a GAN investigation was conducted by the China University of Geosciences to enhance *Pix2Pix* (called *ePix2Pix*) classifications of remote sensing images [14]. They claim improvements to *Pix2Pix* that provide higher classification accuracy when compared to traditional methods. Previous studies reveal that traditional *Pix2Pix* has a limited ability to learn complex image features, such as complicated patterns, thus leading to low classification accuracy among other prediction complications [14, 15]. *ePix2Pix* proposed adding a *Controller* to the model – which now consists of three parts; Generator, Discriminator, and Controller. The Controller allows a relationship between classification and reconstruction, an additional step to improve classification accuracy. Experimental results report that *ePix2Pix* scored higher compared to *Support Vector Machines* (SVM), *Artificial Neural Networks* (ANN), *CycleGAN*, and *Pix2Pix*.

Also, in 2018, Lebedev et al. carried out an experiment on conditional adversarial networks to detect changes in season-varying remote sensing images [16]. Their approach presented three types of tests on synthetic and actual images [16]. In their approach, the Discriminator required three input images to perform the classification (two images for comparison and one for the difference map) - otherwise, the network structure is the same as the *Pix2Pix* structure. Even though the proposed methodology delivered accurate results, changes to *mutable* objects (e.g., vehicles) were also identified as a change in the map (Fig. 6). Although mutable objects should not be considered as a "change in a map," this idea can be utilised to detect changes to immutable objects as well (e.g., buildings).

More recently, Albrecht et al. (2020) presented a method to programmatically identify outdated map regions from current OSM data [13]. This work introduced the use of GAN's, namely *Feature-Weighted CycleGAN* (fw-CycleGAN), to identify any changes in a given geographic area. Although this approach focused on finding changed areas to produce a heat map, it does not explicitly identify/obtain the changed objects themselves, such as a specific newly built building, along with its related ground coordinates. Instead, the objective was to train *fw-CycleGAN* to recognise styles (colour/texture/etc.) in a given set of images and generate OSM-like maps with the same *look* (style/pattern) as output.

[1] https://github.com/phillipi/pix2pix.

3 OSM-GAN Methodology

Our novel OSM-GAN approach integrates many incremental improvements noted in the above strategies for detecting changes in high-resolution spatial images. For example, it targets changes to immutable objects only (i.e., buildings) within a user-defined geographic area digitised on an aerial/satellite image. The designated polygonal Area of Interest (AoI) is then reduced to its minimum bounding rectangle (MBR) coordinates to facilitate further processing operations. We consider the appearance of a new building object or the disappearance of an old object as a change when compared to the current state of the OSM database. For example, if a new building appears in a raster satellite image but is not evident in the OSM vector database, that building is considered a potential changed object within the given AoI. The following system architecture diagram (Fig. 2) illustrates the overall workflow of our automated OSM-GAN approach, and the following sections describe each step in more detail.

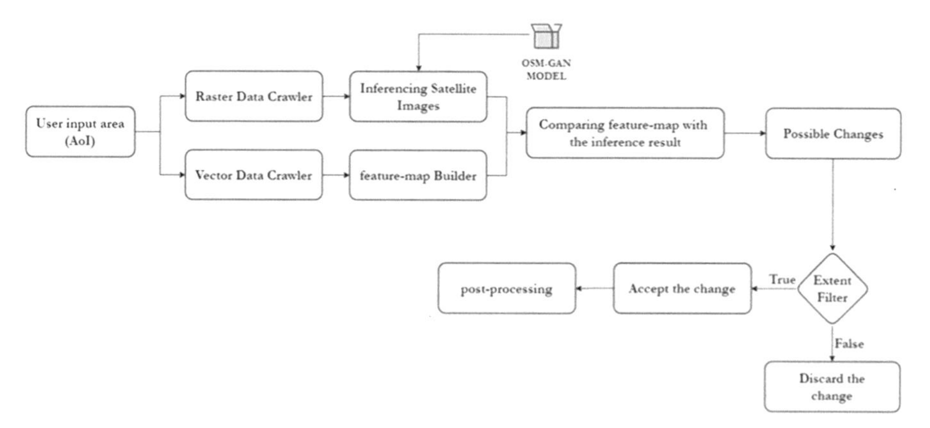

Fig. 2. System architecture diagram of proposed OSM-GAN approach. First, an Area of Interest (AoI) is manually digitised to begin the change detection process. Then, raster and vector data crawling processes launch automatically to download relevant satellite/OSM data. The feature-map prediction phase then starts by using a pre-trained OSM-GAN model. Simultaneously, an OSM feature-map containing predicted buildings is generated. Once both processes complete, feature-map comparisons (OSM to OSM-GAN), extent filtering, and post-processing is applied to each predicted feature-map.

3.1 Spatial Data Crawling and Processing

Data is the most valuable resource for any Machine Learning task, and our AI application requires several spatial data sources for input. As a result, specialised data-mining programs (raster/vector data *crawlers*) were developed to address our spatial data needs. The below sections describe how our necessary data requirements are fulfilled, with further data processing steps described thereafter.

Crawling Vector Data. *OpenStreetMap* [17] is the primary vector data source for this work since OSM vector data (e.g., building footprints) in *GeoJSON* format (*key*:*value* pairs) is relatively easy to handle by post-processing programs [18]. The vector data crawler starts by parsing an OSM *Overpass* query - "a read-only API that serves up custom selected parts of the OSM map data"[2] - into the program. One practical advantage of using the Overpass query is an unlimited freedom to change the OSM feature type (e.g., building, road, river, etc.), the geographic area, and many other OSM attributes by just updating the query without updating the code. The OSM feature extractor and mining program downloads and saves the vector data (within the AoI) into GeoJSON formatted files.

The main contribution of this OSM data is to create building object *mask* images (called *feature-maps*) for use in the training process. Reducing the effect of shadows is another partial benefit of using this object-mask method - since shadows of buildings can affect Mask-RCNN and other traditional image processing techniques. *Pix2Pix* predictions trained using object masks show that the effect of shadows does not significantly affect the accuracy of the OSM-GAN change detection mechanism.

Initially, all OSM crawled data is stored in one large GeoJSON file containing all extracted objects. The main disadvantage of this single file setup is the difficulty of extracting relevant building objects afterwards. Therefore, this large file gets automatically separated into a "one-object-one-file" format and subsequently stored as many individual building object files. This modification results in a significant acceleration of subsequent processes.

To use GeoJSON objects effectively, a translating program first converts them into binary images and stores them in separate directories based on their ground coordinates. Once this process completes, a merging process overlays each of these masks into a single 256×256 pixel sized *feature-map* containing all buildings and used for generating the training images. Figure 3 illustrates an example of separated building objects and the result of the merging process. Note that the white *blobs* in the figure indicate the relevant building objects converted to raster from OSM vector data.

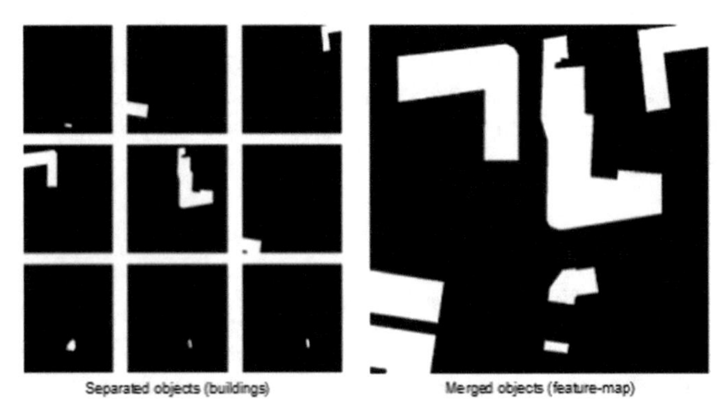

Fig. 3. An example of separated building objects and merged objects. The separated objects were created from the *GeoJSON* data crawled from OSM using their *Overpass* API.

[2] https://wiki.openstreetmap.org/wiki/Overpass_API.

Crawling Raster Data. A raster data crawler was developed for downloading the most up-to-date, freely available satellite imagery related to the same *Overpass* query AoI. Once the vector crawling process completes, raster data mining initiates. There are two main raster mining sub-programs: 1) *Bing* imagery crawler; 2) *Google* imagery crawler – both instructed to apply the relevant crawler at the resolution required for a given task. In this experiment, we chose *Google Earth* images due to their higher quality 30 cm pixel resolution. Also, the downloaded images get automatically cropped (into 256 × 256 pixel tiles), stored, and indexed in a quadtree-based directory structure to make them easier to process in subsequent phases.

Combining and Filtering Input Data. Once the data crawling programs complete successfully, a data processing phase begins to assemble the acquired data according to the requirements of the Deep Learning algorithm. The OSM-GAN training process requires two input images - a satellite image and its conjugate OSM generated *feature-map* image. As the *Pix2Pix* program is pre-configured to use 600 × 300 pixel input images, both the satellite image and its conjugate feature-map are re-scaled (using *OpenCV*[3]) into 300 × 300 pixel tiles and joined together - resulting in the overall 600 × 300 pixel training image sample shown in Fig. 1. However, it was found that low object-dense training images can increase the number of false positives predicted by the model. Therefore, feature-map images that do not contain objects present in their conjugate satellite image are eliminated from the training phase to achieve a higher *Pix2Pix* prediction accuracy.

The Python *NumPy*[4] module is used to determine the number of white pixels, and a ratio is calculated from both the total number of pixels and the number of white (object) pixels. Then a threshold phase determines if the given feature-mask is eligible for use in the training process. Currently, predefined (trial/error) threshold values of 0.25 (lower) and 0.75 (upper) are used, but adaptive thresholds are planned for in the next phase of development.

3.2 Training the OSM-GAN Model

After the above raster/vector data crawling and processing steps, the filtered data is forwarded to the training stage. The *PyTorch* version of the *Pix2Pix* implementation [1] is used to build the OSM-GAN model. The filtered data samples get split into a 3:2 ratio (train:validation), and fed into the *Pix2Pix* algorithm. The OSM-GAN model was trained using an NVIDIA Ge-Force RTX 2060 GPU with CUDA. With this configuration, the training process required 6 h to complete 400 epochs. Figure 4 illustrates the data flow diagram up to this stage.

It was observed that the accuracy of predictions for OSM-GAN models increased with higher resolution imagery – but only up to a certain zoom level, after which prediction accuracy starts to decrease. A series of experiments to discover the best image resolution for training OSM-GAN models revealed that prediction accuracy begins to

[3] https://docs.opencv.org/4.5.2/.
[4] https://numpy.org/.

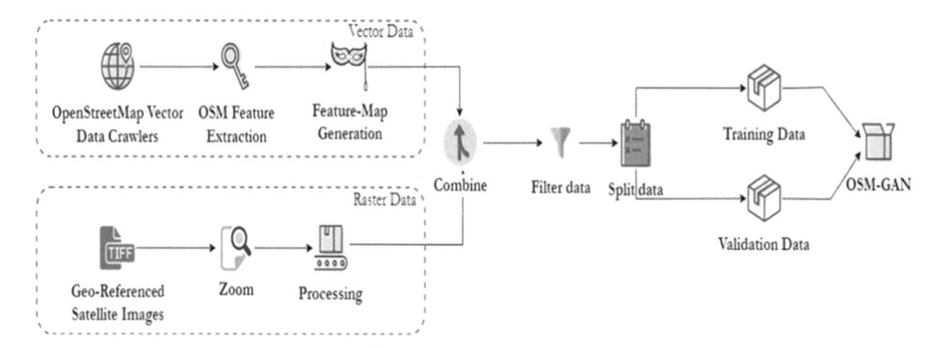

Fig. 4. The fully automated input/output data pipeline for the OSM-GAN framework.

decrease after 0.14m/pixel (zoom level 20) resolution. For example, it was found that OSM-GAN predictions on a 10 cm resolution dataset classified water bodies and fields as buildings. As such, 30 cm resolution images were committed to the qualitative phase of OSM-GAN model predictions. Figure 5 below illustrates some intermediate outputs from the training phase.

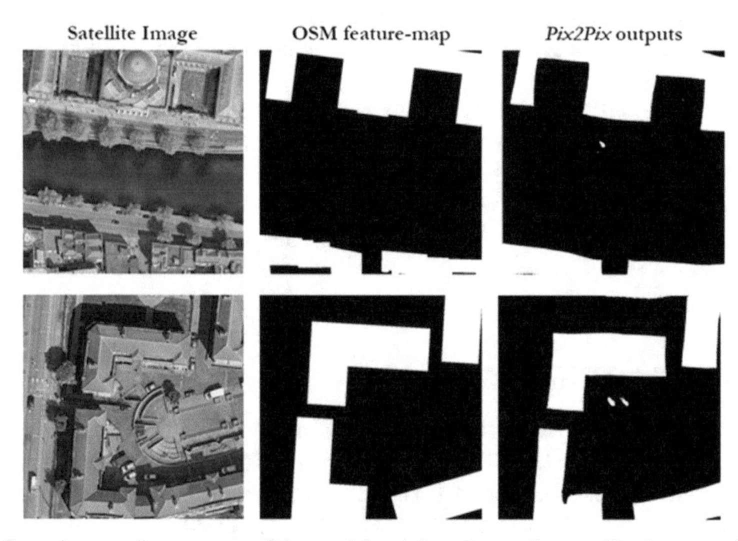

Fig. 5. Some intermediate outputs of the model training phase using satellite images with 30 cm resolution. Note that shadows, water, and vehicles are not classified as buildings by this OSM-GAN approach

3.3 Detecting Changes

The process of detecting actual object changes in the resulting images is relatively straightforward when compared to previous steps. Once the *Pix2Pix* prediction is performed on a given satellite image, the current raster view of the predicted image is reconstructed using the current raster data converted from OSM. Figure 6 compares the satellite image, current OSM data, and the prediction image.

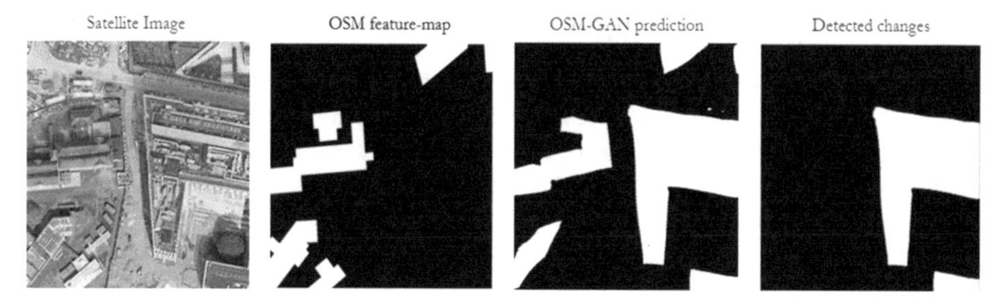

Fig. 6. The OSM-GAN prediction detects building object changes in a spatial image.

Next, the separate building objects are extracted from the prediction results using a conventional contour finding algorithm to perform this operation. After filtering and aggregating the changed object references, Fig. 6 gets extended to include the last image showing detected building changes.

4 Experimental Results

After generating the final image of detected building changes, a post-processing mechanism activates to enhance the overall shape of the changed object(s). First, a *regularisation* operation to reduce the number of vertices is applied to individual building objects to smooth/straighten building outlines. The regularisation phase is a critical step because the OSM code of conduct [19] requires that any objects input to the OSM database must contain a minimal number of nodes. A *perpendicular distance* algorithm was found to work best with our OSM-GAN predictions. For example, Fig. 7 compares a non-regularised polygon to a regularised polygon using perpendicular distance simplification.

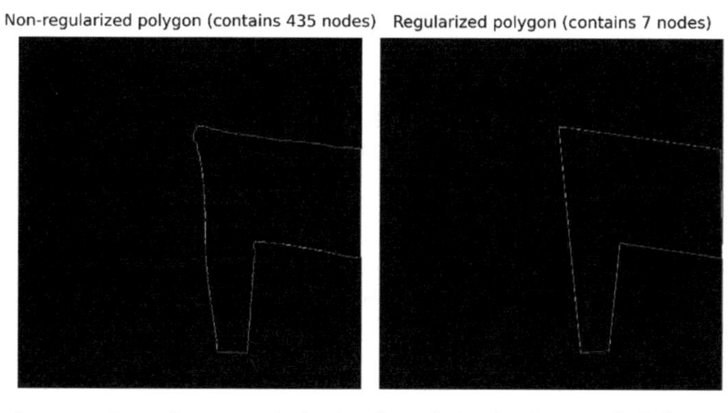

Fig. 7. A comparison of non-regularised and regularised polygons (building outlines).

As described in Sect. 3, OSM building footprint data was utilised to create training datasets for OSM-GAN. One advantage of using an OSM data-based approach is that it can also be used to calculate a confusion matrix (true/false positives/negatives) to evaluate the accuracy of predicted images quantitatively - based on the Object Overlap Matrix and corresponding OSM object labels (used as ground truth). Table 1 shows the experimental results of our OSM-GAN approach for detecting changes to buildings in 30 cm spatial images are at least 88% accurate. However, how to quantitatively evaluate a generated (synthesised) image is still an open and complex problem. In future analyses, we propose to incorporate the "Inception Score" [20] and "Frechet Inception Distance" [21] measures into this metric.

Table 1. Analysis of OSM-GAN prediction accuracy

Accuracy	**88.4%**
Recall	62.0%
Precision	80.5%
F1 score	76.6%

5 Conclusions

This paper proposes an improved solution for detecting changes in spatial images based on combining the incremental developments found in previous work. The overall OSM-GAN system integrates raster/vector data crawling functionality with several other image processing operations, such as image-to-image translation, image difference calculation, and vector to raster conversions into a unified end-to-end workflow. The next phase of research plans to compare change detection results of various experimental combinations of 24-bit satellite images vs 8-bit (grayscale) imagery when applied to OSi[5] and OSM building footprints. Additionally, future experiments will be conducted using Kay[6] - Ireland's national supercomputer for academic researchers. Kay consists of a cluster of 336 nodes, each node having 2 × 20-core 2.4 GHz Intel Xeon Gold 6148 processors; an enormous advantage for minimising model training times.

Acknowledgments. The authors wish to thank all contributors involved with the OpenStreetMap project. This research is funded by Technological University Dublin College of Arts and Tourism, SEED FUNDING INITIATIVE 2019–2020.

References

1. Isola, P., Zhu, J.Y., Zhou, T., Efros, A.A.: Image-to-image translation with conditional adversarial networks. In: Proceedings of the IEEE Conference on Computer Vision and Pattern Recognition, pp. 1125–1134 (2017)

[5] https://www.osi.ie/about/open-data/.
[6] https://www.ichec.ie/about/infrastructure/kay.

2. Goodfellow, I.J., et al.: Generative adversarial networks. arXiv preprint arXiv:1406.2661 (2014)
3. Saqur, R., Vivona, S.: CapsGAN: using dynamic routing for generative adversarial networks. In: Arai, K., Kapoor, S. (eds.) CVC 2019. AISC, vol. 944, pp. 511–525. Springer, Cham (2020). https://doi.org/10.1007/978-3-030-17798-0_41
4. Engel, J., Agrawal, K.K., Chen, S., Gulrajani, I., Donahue, C., Roberts, A.: GANSynth: adversarial neural audio synthesis. arXiv preprint arXiv:1902.08710 (2019)
5. Park, T., Liu, M.Y., Wang, T.C., Zhu, J.Y.: Semantic image synthesis with spatially-adaptive normalisation. In: Proceedings of the IEEE/CVF Conference on Computer Vision and Pattern Recognition, pp. 2337–2346 (2019)
6. Karras, T., Laine, S., Aila, T.: A style-based generator architecture for generative adversarial networks. In: Proceedings of the IEEE/CVF Conference on Computer Vision and Pattern Recognition, pp. 4401–4410 (2019)
7. Benhenda, M.: ChemGAN challenge for drug discovery: can AI reproduce natural chemical diversity? arXiv preprint arXiv:1708.08227 (2017)
8. Mirza, M., Osindero, S.: Conditional generative adversarial nets. arXiv preprint arXiv:1411. 1784 (2014)
9. Gong, M., Su, L., Jia, M., Chen, W.: Fuzzy clustering with a modified MRF energy function for change detection in synthetic aperture radar images. IEEE Trans. Fuzzy Syst. **22**(1), 98–109 (2013)
10. Yousif, O., Ban, Y.: Improving urban change detection from multitemporal SAR images using PCA-NLM. IEEE Trans. Geosci. Remote Sens. **51**(4), 2032–2041 (2013)
11. de Jong, K.L., Bosman, A.S.: Unsupervised change detection in satellite images using convolutional neural networks. In: 2019 International Joint Conference on Neural Networks (IJCNN), pp. 1–8. IEEE, July 2019
12. Papadomanolaki, M., Verma, S., Vakalopoulou, M., Gupta, S., Karantzalos, K.: Detecting urban changes with recurrent neural networks from multitemporal Sentinel-2 data. In: IGARSS 2019–2019 IEEE International Geoscience and Remote Sensing Symposium, pp. 214–217. IEEE, July 2019
13. Albrecht, C.M., et al.: Change detection from remote sensing to guide OpenStreetMap labeling. ISPRS Int. J. Geo Inf. **9**(7), 427 (2020)
14. Wang, X., Yan, H., Huo, C., Yu, J., Pant, C.: Enhancing Pix2Pix for remote sensing image classification. In: 2018 24th International Conference on Pattern Recognition (ICPR), pp. 2332–2336. IEEE, August 2018
15. Lee, M., Seok, J.: Controllable generative adversarial network. arXiv preprint arXiv:1708. 00598 (2017)
16. Lebedev, M.A., Vizilter, Y.V., Vygolov, O.V., Knyaz, V.A., Rubis, A.Y.: Change detection in remote sensing images using conditional adversarial networks. Int. Arch. Photogramm. Remote Sens. Spat. Inf. Sci. **42**(2) (2018)
17. OpenStreetMap (2021). https://openstreetmap.org/. Accessed 22 Apr 2021
18. Overpass-turbo.eu (2021). https://overpass-turbo.eu/. Accessed 22 Apr 2021
19. Code of conduct: OpenStreetMap (2021). https://wiki.openstreetmap.org/wiki/Code_of_conduct/. Accessed 22 Apr 2021
20. Salimans, T., Goodfellow, I., Zaremba, W., Cheung, V., Radford, A., Chen, X.: Improved techniques for training GANs. In: Advances in Neural Information Processing Systems, vol. 29, pp. 2234–2242 (2016)
21. Nunn, E.J., Khadivi, P., Samavi, S.: Compound Frechet inception distance for quality assessment of GAN created images. arXiv preprint arXiv:2106.08575 (2021)

Urban Geography and Spatial Planning

Analysis of Urban Sprawl and Growth Pattern Using Geospatial Technologies in Megacity, Bangkok, Thailand

Pawinee Iamtrakul(✉), Apinya Padon, and Jirawan Klaylee

Center of Excellence in Urban Mobility Research and Innovation, Faculty of Architecture and Planning, Thammasat University, Pathum Thani, Thailand
iamtrakul@gmail.com

Abstract. Urbanization is a spatial process and involves multiple dimensions of the settlements where jobs and housing are the driving force for urban migration, causing the expansion of urban areas from the center of Bangkok to its vicinity. This study focuses on the changes in physical factors affecting urban expansion by using spatial analysis in term of remote sensing statistical models. Three years of aerial photo and satellite imagery studies (2000, 2010 and 2020) are used to determine urbanization which was input into the study of land-use changes to demonstrate the situation of urban sprawl. The changes in land use pattern of Bangkok and its vicinities during 2000–2020 demonstrates the consequences of urban migration based on Urban Index (UI) which presents urban sprawling of Bangkok to its vicinities. The results of analysis demonstrated that the five adjacent provinces of Bangkok have continued its growth with the higher sprawling rate based on entropy calculations ranging from 2.25 to 2.46 in 2020. The expanding from the capital city to the vicinity provinces has resulted due to the nodes of employment are clustered in the capital city. The settlement in the outskirt areas of Bangkok could help in reduction of living expenses within the urban areas, although require longer travel distance from the vicinities of Bangkok. The findings demonstrate the value of geospatial technologies in urban planning and environmental management to cope with the rapid urban expansion, resulting in urban sprawl for sustainable urban strategies and policies by controlling improper future of land use changes.

Keywords: Land use change · Mega City · Urbanization · Urban sprawling

1 Introduction

The world is currently entering a situation of urbanization which many areas have developed into urbanization. Due to the rapid change in the world's population as of 2017, the United Nations (UN) estimated that the world population is approximately 7.6 billion [1]. When Urbanization is one of the major trends emerging today, by 2030, 61% of the global economy will come from activity in 750 major cities which is about 22% of the world's cities. Moreover, it is one of the factors which will attract more population influx to cities from 50% today to 72% by 2050 [2]. Moreover, considering all the major trends,

S. Bourennane and P. Kubicek (Eds.): ICGDA 2022, LNDECT 143, pp. 109–123, 2022.
https://doi.org/10.1007/978-3-031-08017-3_10

urbanization is not only the phenomenon that is most likely to happen in the future. It is also considered the factor that will affect the world economy the most. That is partly due to technological advancements which will enhance the efficiency of future cities. The concept of urbanization has continued to evolve until now. Especially the balancing of the city system can be seen from cities in developed countries such as Germany that have a relatively balanced urban structure [3]. In the case of Germany, it also consists of several large cities similar to Berlin, spread throughout the country which link to the central district attracts and connects a diverse workforce according to the economic role of each city. The small to medium-sized cities are also assigned to play a key strategy of urban development, e.g., a government city center, business capital, etc. [4]. From the increasing in the population with the changes in the economic and social structure, as a result, the existing urban areas cannot meet the needs of the government, the private sector, and the people to build cities in the direction of supporting industrial production, trade, service, and livelihood efficiently [5]. Urban renewal while controlling sprawling situation is an idea to modify, change and maintain existing buildings and urban areas to limit its sprawl and increasing land-use potential to achieve the development goals of economic, social and environmental protection [6, 7].

However, urbanization demonstrated the proportion of the population moving to the suburbs in the increasing rate [8, 9]. As a result, infrastructures and public services are required to expand accordingly to support life [10]. As a result, the construction and investments of public transport has increased by more than 50% to connect the surrounding suburbs to the city's center comfortably and without restrictions. The idea of urban renewal was reintroduced more widely to limit urban sprawling, e.g., compact city, smart city, and job and housing balance which has been received more attention to urban development strategy. Since the situation has been continued to expand and it requires urban planners in applying spatial data to gain a better understanding of complex decisions and responding to the increasing urbanization of population [11]. The measurement of the magnitude of impact even more cumbersome due to the dynamics of urban sprawl must be captured by using effective tool and provide a comparative solution of growth in different scenario of growth control. Since the living expense of the city center is accordingly too high, most people have no choice due to their affordability. The growing problem of affordability is mostly found in the vicinity area of capital city which the spatial mismatch has resulted to the relocation of residents' housing to start a new family in the suburbs, like Bangkok, where it is necessary to commute to the city center for better career and income [12].

In addition, for economic reasons, the need for a large number of workers is attractive together with the economic situation of rural areas where agricultural income is insufficient to cover expenses, as a result, the large number of rural immigrants come to search for job and opportunities in Bangkok. Bangkok as the primate city represents as the capital of every aspect [13]. When Thailand's development plans for shaping future transportation and population growth was explored, it was found that the National Economic and Social Development Plan from the 1st–3rd (1961–1976) plan focused mainly on the construction of public facilities and utilities in Bangkok (electrical, water supply or transportation, education and public health institutions) and support the establishment of industrial factories by focusing on export products. The 4th–6th National Economic and

Social Development Plan (1977–1991), although there are still continuously focusing on facilities and utilities in Bangkok. It began to have decentralized of the urban development to another regions. It can be noticed from the establishment of the Eastern seaboard of Thailand project which was initiated in the 5th National Economic and Social Development Plan (1982–1986). It was also encouraged the distribution of industrial factories in the surrounding provinces of Bangkok. Furthermore, the direction of development is about to push Thailand towards becoming NICs during the 6th National Economic and Social Development Plan (1987–1991). During this development phase, it is also focused on the develop Bangkok and its vicinities to be the center of development of the country which has become the Bangkok Metropolitan Region (BMR) as a receiving area for migrants. Subsequently, in the 7th National Economic and Social Development Plan (1992–1996), more facilities and utilities were in the perimeter provinces to support the expansion of Bangkok. Consequently, the emergence and expansion of the industrial and housing estate business continuously since this period [14].

Bangkok Metropolitan Region (BMR) represents the total area of Bangkok and its vicinity which consists of the other five metropolitan provinces, namely Samut Prakan, Pathumthani, Samut Sakhon, Nakhon Pathom and Nonthaburi. The BMR has a total area of 7,761.6 km^2 [15]. Notably, the surrounding provinces of the capital city was developed and induced by the pattern of economic and social development from the expansion of Bangkok which was resulted from a transition in development policies. The continued construction of houses, facilities and utilities, and roads in the suburbs to support suburban residents and workers has resulted to the growth in a metropolitan area occurs at the fringes. The clustering of construction both industrial and housing estate business continuously at the "urban fringe" which is increasingly characterized development patterns in the perimeter provinces. In addition, government offices and a number of employment areas were relocated to the outskirt areas of Bangkok which attract the number of migrations of the populations in the other provinces coming to work and live. Thus, real estate investment has consequently increased in higher rate due to housing demand, particularly for those who want housing at an affordable price. That is because houses in the inner and middle areas of Bangkok are costly which is the main reason for urban sprawl.

It can be seen from the evidence of failure to adopt the policy which has contributed to the more dispersed urban development patterns. This phenomenon is driven by multi aspect of demographic, economic, geographic, social and technological factors which includes diversity on preferences for living in low-density areas and willingness to accept longer commuting distances. Up to now, it is almost impossible to separate the districts of Bangkok from the districts of the metropolitan provinces in BMR. Because the contiguous development pattern of the city's expansion in the Bangkok metropolitan area is primarily caused by primate city development. Sprawled development patterns are characterized by further distances among residences, jobs and others which has created low density development. The continuation of migration from the countryside creates a sprawled built environment while increasing per-user costs of providing public infrastructures and services in fragmented areas of low-density. Furthermore, the demand for housing and jobs will lead to more car-dependent cities with less shifting travel demand towards public and non-motorized transport [16]. Therefore, this article aims to identify and examine patterns of urban expansion and land use changes in Bangkok and its vicinities through the spatial pattern of urban sprawl by analyzing from satellite images. Land Use/Land Cover (LULC) maps was employed through the use of Landsat imageries (2000, 2010 and 2020), geospatial and Shannon's entropy techniques to quantify the urban sprawl of different provinces of BMR. However, there is some limitations of this approach which is the availability of cloud-free Landsat images to meet the requirement of analysis method. Since the consistency of the time interval of images used is required to allow for further analysis. Thus, the consideration of the specific time interval for analysis should be considered with the other variables, e.g., climate, urban planning and development, regulations and policies that has resulted to the LULC changes.

2 Literature Review

2.1 Compact City

It is a concept of urban development that focuses on limit the city's expansion in the horizontal, particularly in the suburban area. It requires efficient utilization of the land use in the city center to meet the needs to support the people's way of life in both quality and quantity [16]. Urban planning represents a tool to promote compact urban development that often focuses on clearly defining urban growth and land-use boundaries. Instead, it creates a diverse mix of living, working and public service activities in the inner city. Therefore, the development of land and the provision of public services focuses on developing the city's inner areas to accommodate a large number of residents as well as providing public facilities, utilities and areas to support activities in residents' daily life, e.g., working, studying, recreation, etc. This concept is aimed to meet the needs of people in a distance that does not have to travel in longer distance and encourage people to reduce the usage of private cars and turn to public transport, bicycles and walking as the primary means of traveling in their daily lives [17]. For this reason, Shannon entropy together with Geographic Information System (GIS) tools which has been widely applied to measure of sprawl can be adopted to determine the degree of dispersion or compactness of changes in built up area [18]. The urban areas are required to be extracted from the classified images to obtain urban built-up images. The Shannon entropy formula were applied to assess the degree of dispersion or compactness of a spatial variables [18] as shown in Eq. (1).

$$E_n = -\sum_{i}^{n} P_i \log\left(\frac{1}{P_i}\right) \tag{1}$$

where *ln* is the natural log and *Pi* is the value of the geospatial variable in each zone divided by the total land area in that zone. The calculation of Shannon entropy value ranges from 0 to Log(n) which the dispersed distribution is closer to Log(n).

2.2 Jobs-Housing Balance

Balancing the number of homes and work destinations is one of the fundamental concepts of urban management. By this concept, it points out that a balanced city should have a variety built-form to support multi social characteristics. Everyone deserves equal opportunities to live and work in the same city. The main principle is that people can live and work in the appropriate travel distance and time. A more compact development pattern can help reducing car travel and distance while increasing cycling and walking travel. Assessing the balance of most cities, the number of jobs is often compared to the number of homes, which conveys the opportunity to find suitable housing choices for the needs of working people [17]. Boundaries to urban development may be effective in limit the urban growth which can shape its pattern while coping with an increasing in the city's urban population. The expansion has brought pressure on the land cover causing environmental problems. For this reason, this article attempts to assess the impact of urban sprawl on the land use and land cover by focusing on the case of BMR.

3 Methodology

3.1 Study Area

The study of BMR areas consists of 6 provinces as depicted in Fig. 1 which are Nakhon-pathom, Pathumthani, Nonthaburi, Samut Prakan, Samut Sakhon and Bangkok. The expansion both built up area and settlement can be analyzed and the process of analysis is explained in the next section.

Fig. 1. Study area of Bangkok Metropolitan Region (BMR).

3.2 Data Analysis

The satellite image of Landsat 5 and Landsat 8 was applied for classification and analysis of urbanization and land use changes in the BMR area as shown in Table 1.

Table 1. Landsat data (2000, 2010, 2020)

Data type	Date acquired	Season	Path/Low	Reference
Landsat 5 TM	19/01/2000	Winter	129/50 and 129/51	USGS
Landsat 5 TM	17/11/2010	Winter		USGS
Landsat 8 OLI/TIRS	17/12/2020	Winter		USGS

3.3 Data Analysis Process

The preliminary study process is shown in Fig. 2.

- Satellite images of the Landsat-5 TM system and the Landsat-8 OLI system were used as the input data. The TIRS Path/Row 129/50 and 129/51 systems were obtained from the Geo-Informatics and Space Technology Development Agency (Public Organization) and the website of the US Geological Survey (USGS) [19] by selecting images during the dry season and without clouds.
- Geometry correction for Landsat satellite image correction and directional accuracy can be referred to the RAD's 1:50,000 scale of topographic map data. Then, the Ground Control Point (GCP) was set across the image and select a point that is clearly visible on both the satellite imagery data and the topographic maps. It is a process of an image adjustment between the initial image data and geographic coordinates. Geometry can be interpolated by using the formula of linear equations to identify the new coordinates which is called Spatial Interpolation.

- Satellite images can be categorized by using object-based image classification and display satellite images of Bands 4, 5 and 3 (red, green, blue) through selection of a segmentation algorithm. For multi-resolution (Segmentation), the parameters used is scale 30, shape 0.3 and compactness 0.4.
- In this study, it is to determine the land use classification data set and categorize the data set. Land use and land cover were divided into seven categories which are residential and building, agricultural area, forest area, aquaculture area, vacant, water area, and other areas.
- Land use and land cover changes were analyzed by using the Land Change Modeler (LCM), part of the IDRISI Selva program which is a model for analyze potential changes and forecast changes. Results can be analyzed from land use and land cover data for two time periods.

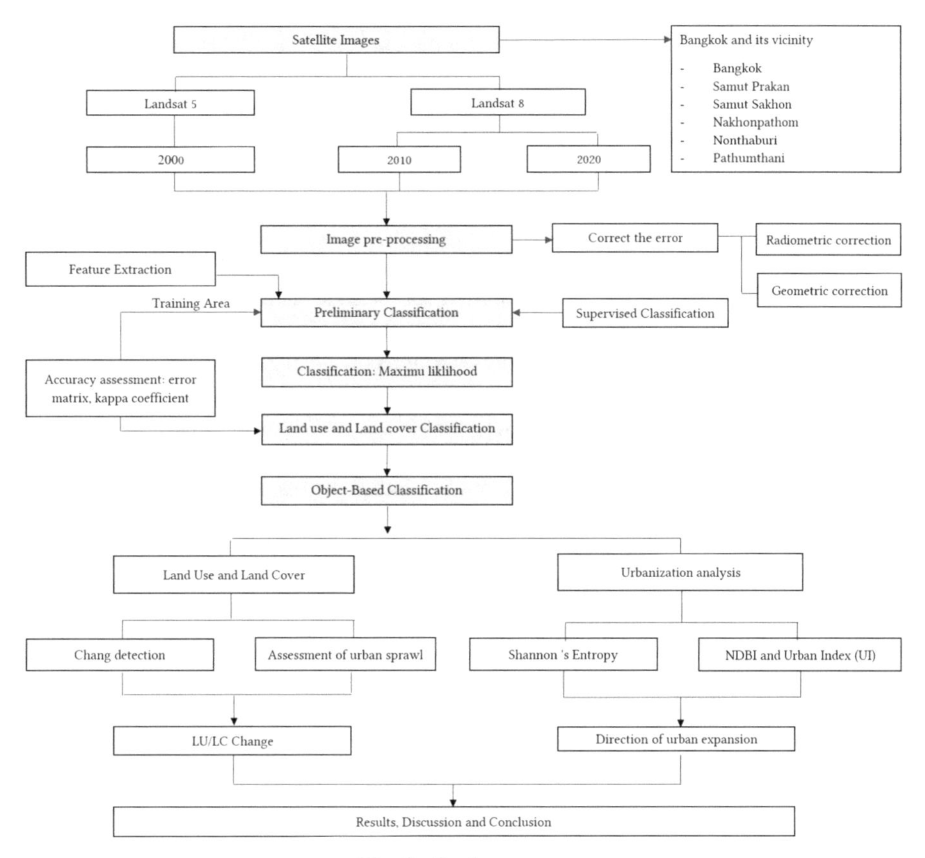

Fig. 2. Study process.

4 Results of Analysis

From the study of urban expansion of BMR, the result of analysis can be used to understand the current problems that cause urban expansion as follows:

4.1 Land Use Change from 2000 to 2020

The land-use change in the suburbs of Bangkok and its vicinities represents the increasing rate. Bangkok is located in the lower Chao Phraya River basin which represents a low level with the former agricultural area. However, over the past 15 years, it was found that the agricultural area around the capital city has been declined slowly. From 2000 to 2010, more than 46% of Bangkok's fertile agricultural land was replaced by urban area. The changes in agricultural area due to urban encroachment and transformed into urban areas. Furthermore, the transformation of urban area has brought the city's growth rate from 2000 to 2020 (Fig. 3) as high as 368%. Without proper land use planning to limit urban growth, suburban farmland was transformed to urbanization, also known as rapid urbanization. In addition, the situation of this urbanization has resulted to an abundance of arable lands, such as areas around Bangkok. It can be noticed the urban growth rate during 2000 and 2010 is about 5.08%. Furthermore, the rate of expansion is increasing to 13.5% during 2010 and 2020. The determination of land use changes in Bangkok and its vicinity (1. Bangkok, 2. Nakhonpathom, 3. Nonthaburi, 4. Pathumthani, 5. Samut Prakan and 6. Samut Sakhon) can be demonstrated in Tables 2, 3 and 4 with the following detail: 1. urban area and built-up area, 2. green area and agricultural area, and 3. other areas.

It was found that the Bangkok area has increased in urban area every year. In 2020, the built-up area of Bangkok was accounted for 39.83% which represented the most urbanized when compared to others. On the other hand, when considering the changes of green area, it was found that in 2000, the province with the highest green area (30.91%) is Nakhonpathom. However, in 2020, the green area of Nakhonpathom (1,712 km^2) had decreased to 28.97% (1,169 km^2). Thus, it is no doubt that the area with the high rate of urban development area in terms of land cover changes is Bangkok and Samut Prakan, respectively, which has consequences from the industrial estate development in these provinces causing migrants to both two provinces. Finally, it has become the challenge of generating sufficient affordable housing to meet demand in total as well as shaping the urban growth pattern to ensure efficient utilization while limit the rate of urban dispersion [20]. Based on the result of analysis in land use changes, it can be noticed that most of the residential area expands along with the transportation network development to the suburban area which represents as ribbon development. The substantial increase in settlement represented by built up area (urban area) indicates the conversion of most of the land cover types into impervious surfaces with urban development. Most of urban area is taking place within the metropolis of Bangkok rather than other provinces. However,

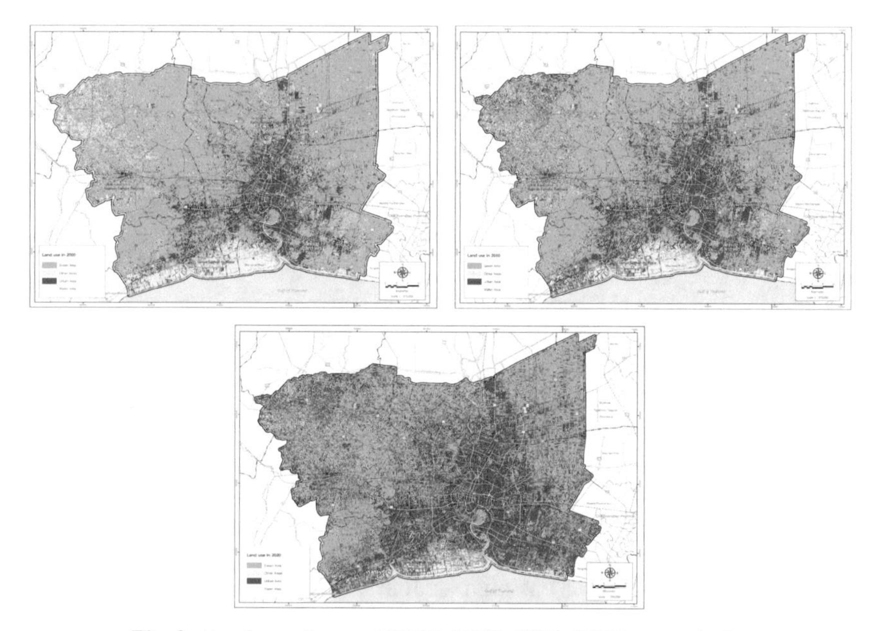

Fig. 3. Land use change of BMR (2000, 2010, 2020) respectively.

an increasing in urban areas in the outer Bangkok area (vicinity provinces) demonstrated the increasing trend of a density of buildings, including land prices that are not as high as the Bangkok area. It requires further step to capture the dimensions of sprawling phenomena happened in BMR.

4.2 Analysis of Factors Affecting Urban Expansion

The analysis of physical factors contributed to urban expansion can be determined based on fundamental factors of urban functions/activities.

Table 2. Land use changes of BMR in 2000

No.	Area changes in 2000 (square kilometers)					
	Green area	Proportion (%)	Urban area	Proportion (%)	Other area	Proportion (%)
1	1,053.78	19.02	408.58	34.00	113.41	11.53
2	1,712.83	30.91	158.29	13.17	264.98	26.94
3	509.59	9.20	91.58	7.62	35.99	3.66
4	1,331.00	24.02	141.59	11.78	45.79	4.65
5	448.92	8.10	249.75	20.78	280.74	28.54
6	484.81	8.75	151.85	12.64	242.72	24.68
Total		100.00	1,201.64	100.00	983.62	100.00

Table 3. Land use changes of BMR in 2010

No.	Area changes in 2010 (square kilometers)					
	Green area	Proportion (%)	Urban area	Proportion (%)	Other area	Proportion (%)
1	561.14	16.87	514.26	40.73	334.94	21.86
2	1,021.11	30.70	108.06	8.56	505.52	33.00
3	349.37	10.50	110.90	8.78	113.23	7.39
4	816.93	24.56	160.55	12.71	353.04	23.05
5	182.29	5.48	249.06	19.72	162.56	10.61
6	395.75	11.90	119.92	9.50	62.60	4.09
Total		100.00	1,262.75	100.00	1,531.89	100.00

Table 4. Land use changes of BMR in 2020

No.	Area changes in 2020 (square kilometers)					
	Green area	Proportion (%)	Urban area	Proportion (%)	Other area	Proportion (%)
1	641.21	15.89	570.82	39.83	247.54	21.11
2	1,169.06	28.97	199.25	13.90	502.38	42.83
3	377.46	9.35	120.16	8.38	89.21	7.61
4	1,084.04	26.86	160.23	11.18	199.18	16.98
5	361.92	8.97	222.45	15.52	86.48	7.37
6	402.23	9.97	160.29	11.18	48.08	4.10
Total		100.00	1,433.20	100.00	1,172.89	100.00

Note: $Growth\ rate_{(2010-2020)} = \frac{(Total\ Urban\ Area_{2020} - Total\ Urban\ Area_{2010})}{Total\ Urban\ Area_{2010}} \times 100 = \frac{(1{,}433.20 - 1{,}262.75)}{1{,}262.75} \times 100 = 13.5$

This study employed the differential analysis of urban physical characteristics to determine the proportion, density, according to urban planning standards in various dimensions based on the Urban Index (UI) analysis. Near Infrared (NIR) and Short Wavelength Infrared (SWIR) [21] are employed to demonstrate the level of the density of the city can be determined as shown in Eq. (2).

$$UI = \frac{SWIR - NIR}{SWIR + NIR} \tag{2}$$

where: UI = Urbanity Index

NIR = Near Infrared. The wavelength ranges between 0.85–0.88 μm.

SWIR = Shortwave infrared. The wavelength ranges between 2.11–2.29 μm.

It was found that compact growth scenarios characterized by high-density settlements had advantages over low-density areas in terms of overall urbanization at any size of city. Therefore, each type of urban model should be considered based on the pattern of urban expansion to the understanding the changes in population in different provinces. This study employed the list of urban factors as shown in Table 5 to determine the level of settlement of BMR. Based on the Urban Index (UI), the rate of urbanization can be determined and demonstrated the high urbanization in Bangkok and Pathumthani.

Table 5. Factor analysis of the urban sprawling

Factors	Definition	Unit	References
Built-up	The urban area is analyzed from satellite images	Sq. km	Asfaw Mohamed & Hailu Worku, 2019; Dandan Liu, Keith C. Clarke & Nengcheng Chen, 2020 [23, 24]
Non-built-up	Other non-urban areas such as green areas water source area other source areas, etc.	Sq. km	Asfaw Mohamed & Hailu Worku, 2019; Maslow, 1943; R.J. Schneider, 2013 [23, 25, 26]
Slope	*Slope level* – 0–5 – 5–10 – 10–21 < (Critical slope) – 21–30	Percent (%)	Asfaw Mohamed & Hailu Worku, 2019; Dandan Liu, Keith C. Clarke & Nengcheng Chen, 2020 [23, 24]
Elevation	1200–1500 1500–1900 0.6 >1900	meter	Asfaw Mohamed & Hailu Worku, 2019; Dandan Liu, Keith C. Clarke & Nengcheng Chen, 2020 [23, 24]
Water area	Area size of water resources in the area from satellite images	Sq. km	Asfaw Mohamed & Hailu Worku, 2019; Maslow, 1943; R.J. Schneider, 2013 [23, 25, 26]
Green area	The size of the green area in the area from the satellite image	Sq. km	Asfaw Mohamed & Hailu Worku, 2019; Maslow, 1943; R.J. Schneider, 2013 [23, 25, 26]
Bare land	Empty/unused areas of satellite imagery	Sq. km	Asfaw Mohamed & Hailu Worku, 2019; Maslow, 1943; R.J. Schneider, 2013 [23, 25, 26]

The expansion of urban areas (Bangkok and its vicinity) can then be determined together with the ratio of city area to total area (%) to observe the pattern of changes. Furthermore, net immigration rate (per 1,000 population) was applied to understand the movement of population [18]. This is to compare the changes of built-up area together with population growth, the average move-in-rate was adopted and calculated by using Net Migration Rate (NMR). NMR is defined as *"the number of net migrants* $(I - O)$ *per mid-year population* (P) *per 1000 population" where* $I =$ *Total number of in-migrants;* $O =$ *total number of out-migrants;* $P =$ *Mid-year population size* [22]. Finally, the calculation of the Shannon entropy value was utilized to assess the pattern of urban distribution and confirm the dispersion of the urban growth. The level of dispersion or compactness of the spatial expansion of different provinces can be calculated and the result of urban expansion pattern can be summarized as shown in Table 6. Based on the analysis of the land cover maps of 2000, 2010, and 2020 to identify built-up and non-built-up area, it can input for further process of urban sprawl measurement by using Shannon's entropy.

Table 6. The expansion of urban areas in Bangkok and its vicinity

Province	Area (Sq. km.)	Urban area (Sq. km.)	Non-urban area (Sq. km.)	Average move-in rate (no. of people)	Ratio of city area to total area (%)	Net immigration rate (per 1,000 pop. in 2019)
Bangkok	1,576	897	679	36,874	56.92	−11.19
Pathumthani	1,811	766	1,045	41,487	42.30	2.24
Nakhonpathom	2,042	473	1,569	13,015	23.16	0.80
Nonthaburi	636	265	371	27,307	41.67	2.65
Samut Prakan	919	385	534	26,817	41.89	1.34
Samut Sakhon	879	255	624	8,868	29.01	0.09

Note. Areas with high urbanization proportions in both provinces were selected as sample areas for analysis

* Database from satellite image analysis (USGS, 2020). (Accuracy of analysis = 88.4%)

* Net immigration rate, from $MNR = (I - O)/P \times 1000$(Number of In-immigrants (I), Number of Out-immigrants (O), Mid-year population (P)) and Ratio of city area to total area = A/B*100 (a = Area, B = Urban area)

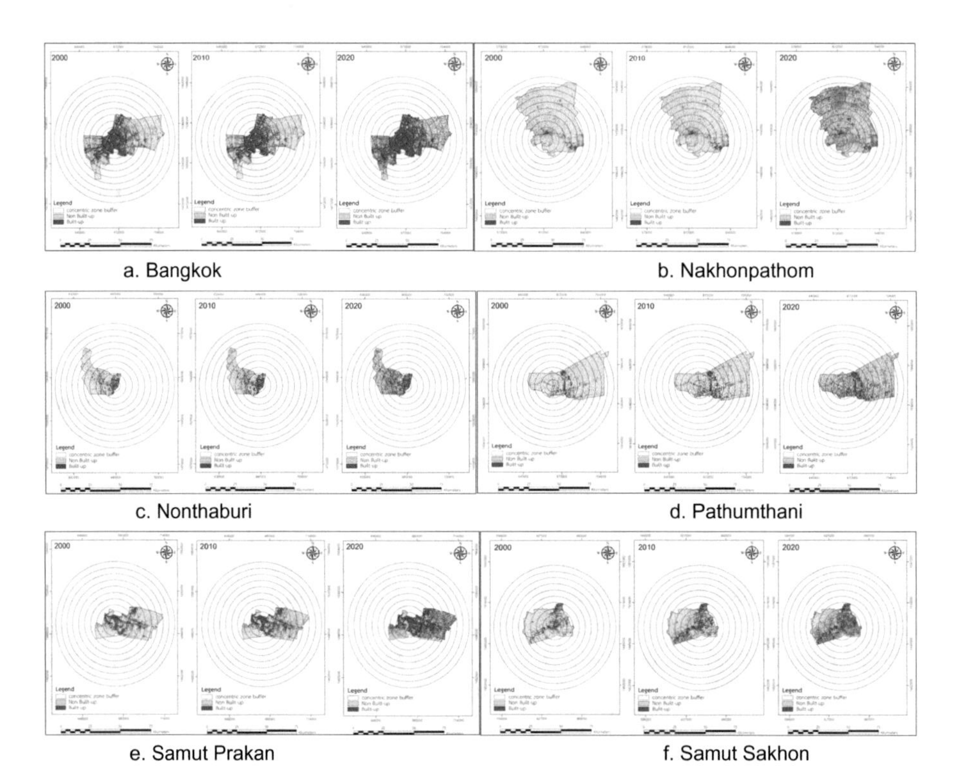

Fig. 4. Reclassified maps showing built-up and non-built-up areas for the years; 2000, 2010, and 2020.

Table 7. Values of relative and Shannon's entropy

Province/Year	Built up area (km^2)			Value of Shannon's entropy			Value of relative Shannon's entropy		
	2000	2010	2020	2000	2010	2020	2000	2010	2020
Bangkok	408.58	514.26	570.82	2.52	2.44	2.25	0.84	0.79	0.75
Nakhonpathom	158.29	168.06	199.25	2.5	2.46	2.43	0.82	0.86	0.92
Nonthaburi	91.58	110.9	120.16	2.46	2.42	2.4	0.84	0.88	0.94
Pathumthani	141.59	160.23	160.55	2.48	2.46	2.46	0.79	0.86	0.92
Samut Prakan	222.45	249.06	249.75	2.5	2.46	2.46	0.82	0.86	0.9
Samut Sakhon	119.92	151.85	160.29	2.45	2.42	2.4	0.8	0.84	0.89
BMR	1142.41	1354.36	1460.82	2.45	2.43	2.42	0.82	0.84	0.87
Log (15)				2.644					

The analysis of urban sprawl based on its spatial forms of different provinces would contribute to prioritizing policies and specific regulations in dealing with the dominant form of different areas which was demonstrated in Fig. 4 and Table 7.

5 Conclusions

This article applied GIS and remote sensing techniques to examine the land use changes which took place in BMR from 2000 to 2020.While the population was increasing, the situation in land-use changes of the BMR can be determined its sprawling pattern by using the Shannon's entropy. It was found that Bangkok has a negative overall net migration rate of −11.19% (Table 7), indicating that the migration trend of urban residents has a large number of people moving away from Bangkok. Based on such a situation, it is possible that most of migration searching for the affordable price of housing with more choices of housing estates in the outer area of the metropolitan. Furthermore, there is also the development of road systems and mass transit connectivity represent as a competitive mode choice for suburban commuters. Consequently, it has resulted to migrations of the populations to the suburbs and vicinities of Bangkok and causing the migration rate of the other five adjacent provinces of Bangkok to be positive as shown in Table 6.

This study also attempts to measure the urban sprawl by using Shannon's entropy which was calculated based on primary spatial forms of urban sprawl [18, 22, 27]. The obtained Shannon's entropy values indicated that spatial expansion of urban city is sprawling more from year 2000 to 2020. The result of analysis in Table 7 demonstrates that for the twenty (20) years, there has been a considerable change in the entropy values. Log (15), which is 2.64 is the highest Shannon entropy value, and 2.25, 2.43, 2.40, 2.46, 2.46 and 2.40 are respectively Shannon's entropy values for Bangkok, Nakhonpathom, Nonthaburi, Pathumthani, Samut Prakan and Samut Sakhon in 2020. From the relative Shannon entropy in for the three periods, (from the value of 0 to 1, minimum to maximum), it is observed that all provinces present the increasing value except Bangkok. The relative Shannon entropy in the year 2020, the values 0.75, 0.92, 0.94, 0.92, 0.90 and

0.89 were high for all provinces, except Bangkok. The minimum value of 0.75 was for Bangkok and the maximum value of 0.94 was for Nonthaburi. These values are closer to 1 for all of five adjacent provinces of Bangkok indicates that the suburbs are sprawling. It is clearly confirmed that the findings represents greater importance to call for appropriate policies and sustainable urban planning to allocate proper preparation and utilize the historical geographical data set efficiently [28].

Acknowledgments. This project is funded by the National Research Council of Thailand (NRCT) under the project entitled "The Study of Factors Influence on an Inequality of Urban Mobility towards Sustainable Development: Case Study of Bangkok and Its Vicinity", contract no. NRCT/757/2563. It is also conducted under the research unit which is supported by the Faculty of Architecture and Planning Research Fund, Thammasat University, contract no. TDS 10/2021, and partially supported by Center of Excellence in Urban Mobility Research and Innovation, Faculty of Architecture and Planning, Thammasat University, Pathumthani, Thailand.

References

1. United Nations: Department of Economic and Social Affairs, Population Division. World Population Prospects 2017 – Data Booklet (ST/ESA/SER.A/401) (2019)
2. PricewaterhouseCoopers LLP: Five megatrends and their implications for global defense & security, November 2016 (2019)
3. United Nations, Department of Economic and Social Affairs: Population Division: World Urbanization Prospects: The 2018 Revision (ST/ESA/SER.A/420). United Nation, New York (2019)
4. United Nations, Department of Economic and Social Affairs, Population Division. The World's Cities in 2016 – Data Booklet (ST/ESA/SER.A/392) (2016)
5. Fitzpatrick-Lins, K.: Comparison of sampling procedures and data analysis for a land-use and land cover map. Photogramm. Eng. Remote. Sens. **47**(3), 343–351 (1981)
6. Turner II, B.L., Skole, D., Sanderson, S., Fischer, G., Fresco, L., Leemans, R.: Land use land cover change. Science/research plan (IGBP Report No. 35 and HDP Report No. 7) (1995)
7. Veldkamp, A., Lambin, E.F.: Predicting land-use change. Agric. Ecosyst. Environ. **85**, 1–6 (2001)
8. Iamtrakul, P., Srivanit, M., Klaylee, J.: Resilience in urban transport towards hybrid canal-rail connectivity linking Bangkok's canal networks to mass rapid transit lines. Int. J. Build. Urban Interior Landscape Technol. (BUILT) **10**, 27–42 (2018)
9. Labs, C.: IDRISI Spotlight: The Land Change Modeler. Clark University, Worcester (2013)
10. Araya, H.Y., Pedro, C.: Analysis and modeling of urban land cover change in Setúbal and Sesimbra, Portugal. Remote Sens. **2**, 1549–1563 (2010)
11. Bollens, J.C., Schmandt, H.J.: The Metropolis: Its People Politics and Economic. Life Harper Row Publishers, New York (1965)
12. Apinya, P., Iamtrakul, P., Thanapirom, C.: The study of urbanization effect on the land use changes and urban infrastructures development in the Metropolitan Areas, Thailand. IOP Conf. Ser. Earth Environ. Sci. **738**, 012077 (2021)
13. D'Antonio, M., Colaizzo, R., Leonello, G.: Mezzogiorno/Centre-North: a two-region model for the Italian economy. J. Policy Model. **10**, 437–451 (1988)
14. Iamtrakul, P., Satichob, P., Hokao, K.: Comparing the efficiency of urban components in proximity to transit service area. Int. J. Build. Urban Interior Landscape Technol. (BUILT) **2**, 21–34 (2018)

15. Iamtrakul, P., Raungratanaamporn, I.: The Study on promoting hybrid canal rail connectivity in Bangkok and its vicinity. Int. J. Build. Urban Interior Landscape Technol. (BUILT) **8**, 13–26 (2018)
16. OECD: Rethinking Urban Sprawl: Moving Towards Sustainable Cities. OECD Publishing, Paris (2018)
17. Bhatta, B.: Analysis of urban growth pattern using remote sensing and GIS: a case study of Kolkata, India. Int. J. Remote Sens. **30**, 4733–4746 (2009)
18. Ernest, B., Ebenezer, B.: Urban sprawl and its impact on land use land cover dynamics of Sekondi-Takoradi metropolitan assembly, Ghana. Environ. Challenges **4**, 100168 (2021)
19. The Geo-Informatics and Space Technology Development Agency. Satellite images of the Landsat (2021)
20. Apinya, P., Benjamin, P., Iamtrakul, P., Klaylee, J.: The study on association between urban green space and temperature changes in Mega City. In: 2020 International Conference and Utility Exhibition on Energy, Environment and Climate Change (ICUE) (2020)
21. Markham, B.: The Landsat 8 satellite, NASA and the U.S. Geological Survey (USGS) (2013)
22. Mosammam, H.M., Nia, J.T., Khani, H., Teymouri, A., Kazemi, M.: Monitoring land use change and measuring urban sprawl based on its spatial forms: the case of Qom city. Egypt. Remote Sens. Space Sci. **20**(1), 103–116 (2017)
23. Mohamed, A., Worku, H.: Quantification and mapping of the spatial landscape pattern and its planning and management implications a case study in Addis Ababa and the surrounding area, Ethiopia. Geology Ecology Landscapes **5**(3), 161–172 (2019)
24. Liu, D., Clarke, K.C., Chen, N.: Integrating spatial nonstationarity into SLEUTH for urban growth modeling: a case study in the Wuhan metropolitan area. Comput. Environ. Urban Syst. **84**, 101545 (2020)
25. Maslow, A.H.: Hierarchy of needs: a theory of human Motivation Kindle Edition. Psychology Classics (1943)
26. Schneider, R.J.: Theory of routine mode choice decisions: an operational framework to increase sustainable transportation. Transp. Policy **25**(C), 128–137 (2013)
27. Tian, L.: Measuring urban sprawl and exploring the role planning plays: a Shanghai case study. Land Use Policy **67**, 426–435 (2017)
28. Klaylee, J., Iamtrakul, P.: Urban planning measures for smart city development. In: The 11th International Structural Engineering and Construction Conference (ISEC-2011), 26–31 July 2021, in Cairo, Egypt (2021)

WEB-GIS for Transportation System of Oran City

Khedidja Belbachir(✉) and Fatima Zohra Belhouari

Center of Space Technics, Oran, Algeria
Khadidjabelbachir@gmail.com

Abstract. The proliferation of geographic information systems has generated growing needs for having integrated vision of the territory, many initiatives are emerging. Their purposes provide application form the distribution of catalogs to the online posting, exchange or even standardization and co-production of spatial data. In this paper, we present the project named "Openstreet Map data quality evaluation using high resolution imagery and reflection for contribution process of contribution in rural areas of Oran city". Our objective is to develop a webGIS platform offering various functions for exchange and update spatial data under client/server architecture.

Keywords: OSM · Territory · Standardization · Spatial data · WEBGIS

1 Introduction

Actually, spatial data and geomatics technics tools are no longer reserved for experts in the field but now affect all stakeholders in the territory (town planners, geologists, foresters, network managers, etc.). Faced with this proliferation of geographic information systems, many partnership initiatives are emerging. Their purposes tend to evolve gradually: form sharing to producing a standardized spatial data.

The term "webmapping/web mapping" is broad because it brings together various skills and technics [1]. This term designates "in the broad sense, everything that comes under online cartography on the Internet." This generic term encompasses different types of cartographic applications, ranging from a simple "viewer" to a thematic mapping tool, or even online GIS. Their common point is that they are accessible through a simple Internet browser. According to [2], there are three different levels of web mapping depending on the functionality it offers the user:

Simple web mapping "viewer" where only data consultation is possible, the visualization being static;

"Dynamic" web mapping means that the map displayed online is interactive. It is possible for a user to use simple functions such as zooming, choosing to display layers or even display tooltips.

This level makes it possible to go beyond the downloading of static maps and to access geographic data contained on a cartographic server, such as for example a database or GIS files [3].

S. Bourennane and P. Kubicek (Eds.): ICGDA 2022, LNDECT 143, pp. 124–132, 2022.
https://doi.org/10.1007/978-3-031-08017-3_11

The GIS-web in addition to the two previous functions, it provides more advanced functions such as the manipulation of geographic entities and carrying out advanced attribute and spatial queries.

In this paper we present the WEBGIS application for transportation system of Oran city "SIG-WEB RRO", allowing the management of the road network of Oran city. with a view to arranging transport on demand. The developed version has been enriched with functionalities meeting the needs of the project entitled: "Evaluation of the quality of OpenstreetMap (osm) data by high-resolution imagery and reflection on the process of contribution in rural areas of 'Oran' city. By an update on OSM data.

2 Project Development

The main objective of our contribution is to develop a web platform for the benefit of project members to be able to exchange and manipulate data, under the Ethernet of our establishment CTS.

The developed platform is a GIS-web that allows users to view, display attributes, add, update geographic features, and doing queries.

As this is a computer tool, the design is almost identical to that of the "web mapping" application. The notable difference is observed in the development monitoring model.

The one in this work is "V" model because of the specifications functions are known from the beginning.

2.1 Platform Architecture

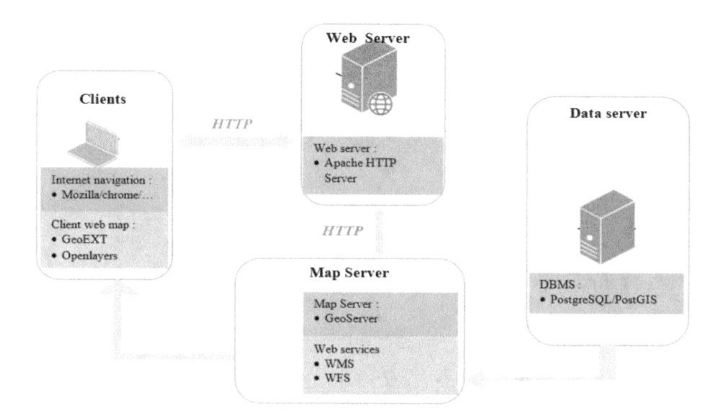

Fig. 1. Client/server architecture

The SIGWEB-RRO application is based on client/server architecture as shown in the Fig. 1 developed using windows as operating system, and includes other free software used for processing data before storage (projection, conversion, ...) styles and symbols creation for WMS maps.

Servals web-mapping tools are available for the development of the web application, we opted for OpenSource solutions, because they allow independence and agility.

Web Services. The online posting and access to spatial data on the internet must comply with formats and protocols defined by the (Open Geospatial Consortium) OGC [7]. In our case, we need services that allow the visualization and the interrogation of spatial data. The Web Map Service (WMS) [7] and the Web Feature Service (WFS) meets this dual objective.

Database Management System. PostgrSQL and MySQL are the tow DBMS most used [4] each one has its own spatial cartridge to manage spatial data.

Map Server. As the most used of open source software map servers we find MapServer [5] and Geoserver [6] booth of them are in accordance with the recommendations of the OGC. The two map servers are equivalent in the possibilities they allow publishing spatial data on the Internet. But GeoServer is simpler to use it has an easy interface. This is why our choice fell on GeoServer as a map server. SIGWEB-RRO platform component are the following:

- A repository of spatial data files (SHP and OSM format)
- Geoserver as a map server for generating geo web services (WMS,WFS)
- PostgeSQL/postgis spatial data base server.
- Apach as a proxy web.
- GeoEXT and Openlayers as map viewers with GIS functions.

2.2 SIGWEB-RRO Component

The developed platform ensures the following operations:

Authentication: SIGWEB-RRO ensures data consistency with the authentication through data privileges manipulation. Authentication has been shown in the Fig. 2 below.

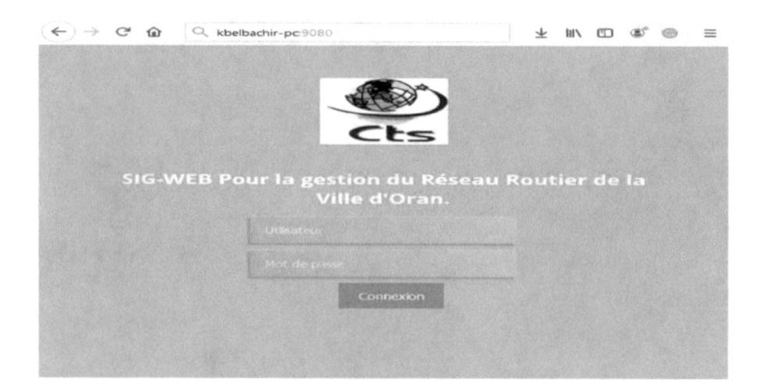

Fig. 2. Authentification page.

Navigation: Display, zoom movement in the map.
Data catalog: access the data catalog and select the layers to display.
Display of attribute data: display of data table of the selected layer.
Display of tooltip: display of attributes on tooltip when clicking on the shapefile.
Update: of both shape and attribute features.
Query: on Postgis database.

2.3 Platform Functionalities

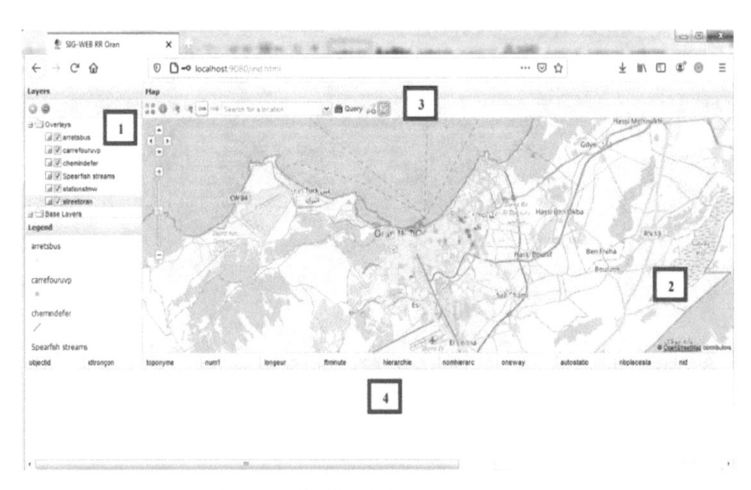

Fig. 3. SIGWEB-RRO interface.

The platform interface is interactive. Its visual appearance is simple and meets perfectly to the specifications of the project. As shown in Fig. 3 the interface is composed of four different areas:

Area 1: Presents the data catalog, the (+) and (–) buttons allow the addition and removal of thematic layers.

Area 2: Map display, it has a navigation tool in the map to perform the zoom in and zoom out.

Area 3: Toolbar grouping the platform's functionalities. The role of each icon is shown in Table 1.

Table 1. Map toolbar.

Icone	Description
	Max zoom out for the map
	Pop-up of selected feature
	zoom in and zoom out
	Legends
Search for a location ...	Geocoder
Query	Spatial query tool
	Data creation
	Delete and update data.

Area 4: Data attribute of the selected or query result layer.

2.4 OSM Data Under the SIG-WEB

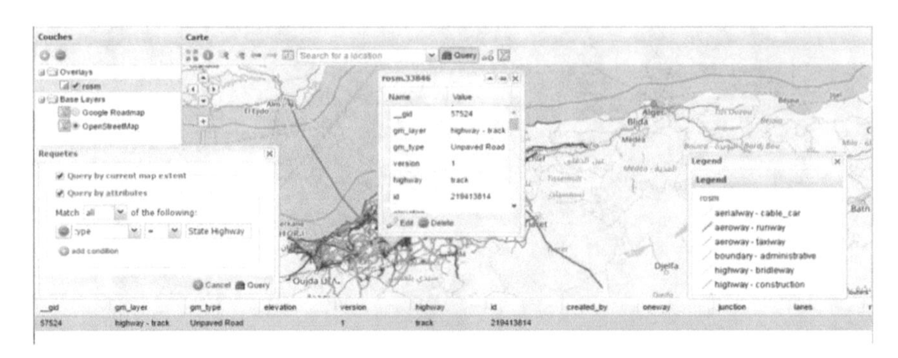

Fig. 4. OSM data.

Figure 4 is a part of the road network of Oran city under the developed GISWEB-RRO. The color representation style was configured under the Desktop GIS tool then imported to the data server then shared through a cartographic server using geographic web services (WMS, WFS).

Layer Adding. The addition of a thematic layer is done by clicking on the (+) button of area 1; the form shown in Fig. 5 explores the data loaded to the "Geoserver".

The window in Fig. 5 is presented in the table with the "Title" presenting the alias of the data and an "id". Let us take as an example the following id "dz: chemindefer"; dz is the name of the workspace on the map server, and "chemindefer" is the name of the table on the data server. The operation developed offers the possibility of loading data or even base maps from remote servers.

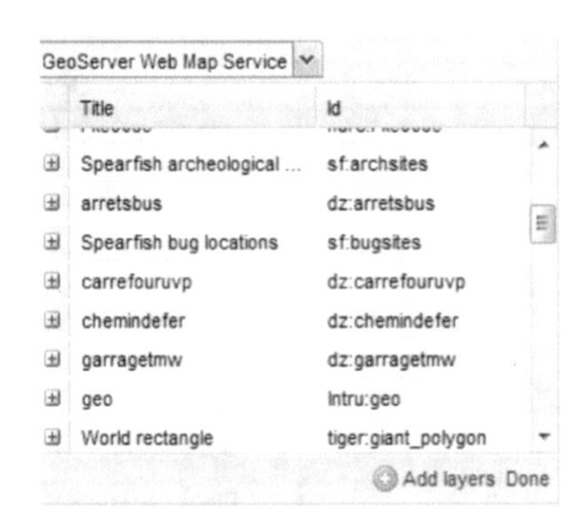

Fig. 5. Available data on the Geoserver

2.5 Creation of Geographic Entities

To create new spatial objects, you must select the data from the catalog. Use the drawing icon on the toolbar for the creation of the desired shape (point, line or polygon). Before saving, you must fill in the attribute fields, as Fig. 6 shows.

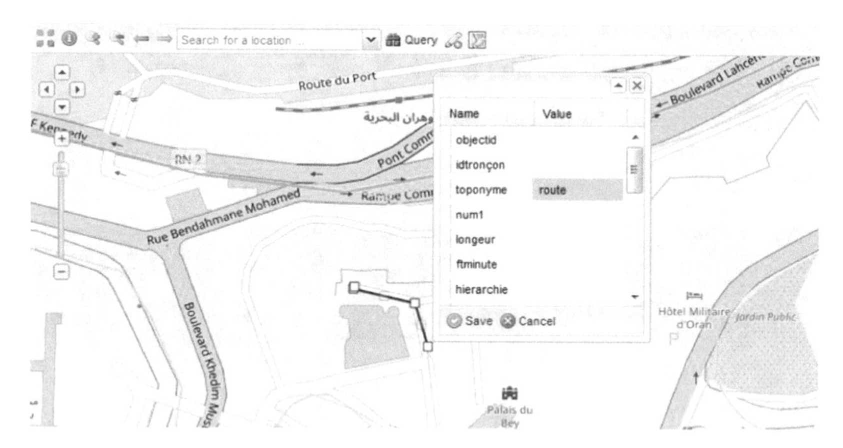

Fig. 6. Geographic object creation.

The creation of new data is not supported by the platform; the operation is only favorable to the selected data.

Updating Spatial Data. The update on the shapes or attributes will be carried out then saved as Fig. 7 shows.

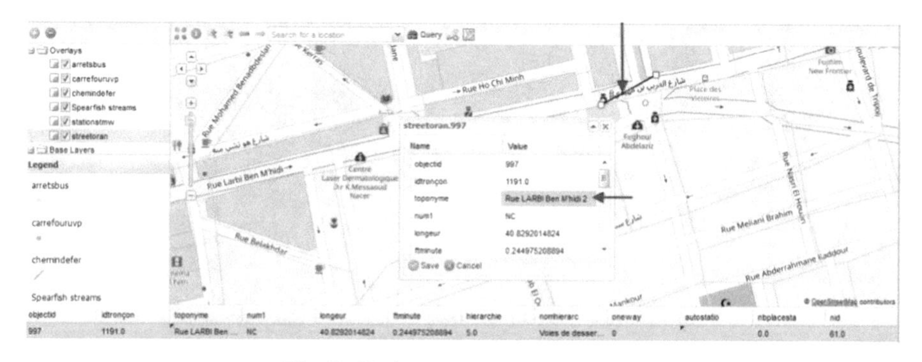

Fig. 7. Update on geographic entities.

The developed system ensures data consistency; this can be verified after validation of the modification indicated for example by arrows in Fig. 7:

– Moving road segment;

– Modification of “toponym” attributes.

To confirm changes an SQL query at the DBMS level is done, the result is shown in Fig. 8.

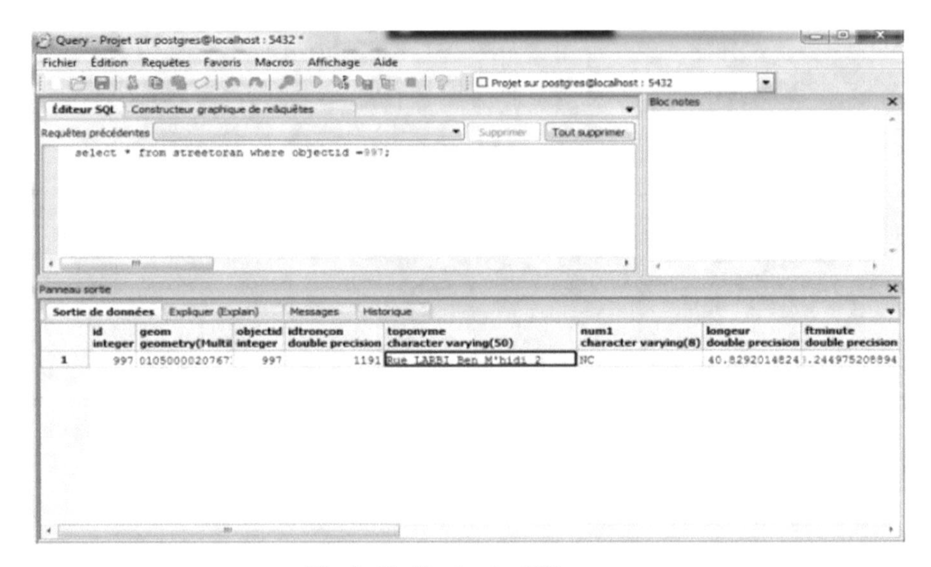

Fig. 8. Verification by SQL query.

The changes made on the developed platform are duplicated on the database.

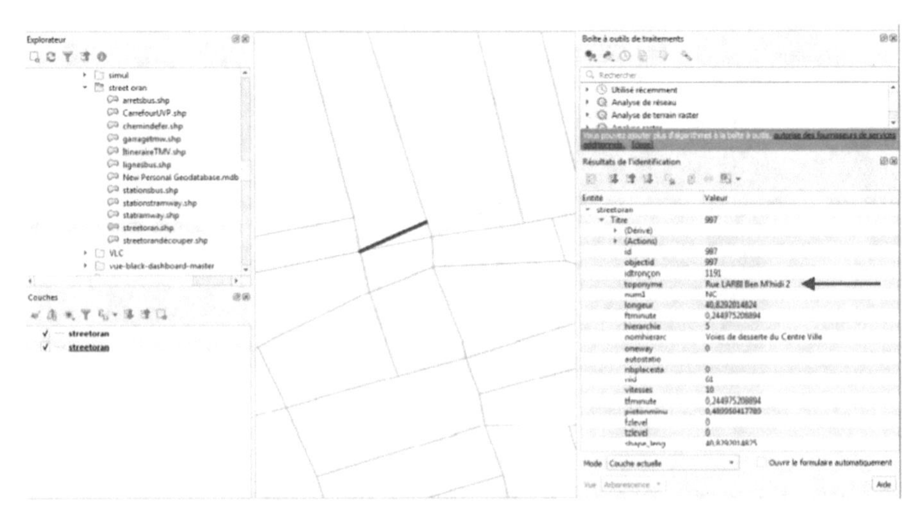

Fig. 9. “Streetoran” entity modified on the map server.

Figure 9 shows a display of the “streetoran” data under the GIS tool. The modifications are listed on the Geoserver. This approves the interoperability of the technical architecture deployed.

2.6 Query

A query can be done on the data selected from the catalog. The example shown in Fig. 10 is a selection of all sections indicating a speed limited equal to 60 km/h.

Fig. 10. Result of a query.

The result of the query is presented in the displaying data area, by clinking on attributes tuple feature is zoomed in, corresponding to the query condition on the map, in bold colored lines with its attribute table in writing.

3 Conclusion

The WebGIS developed within the framework of this project is an operational tool, corresponds to the needs analyzed in the phase of its design. It offers fundamental functionalities of a GIS; by adding different data resources (raster & vector), updating shapes and doing queries on the attribute data. Our application can be a useful technical support for several projects with a public user rank.

The scope of our contribution will be the subject of the creation of a spatial data infrastructure (SDI) very similar to web-GIS but richer in terms of popularizing information and geographic data. This implementation imposes to use and produce normalized spatial data, to strengthen the use of open data.

References

1. Goëta, S.: Rapport sur l’open data, pour le Conseil départemental de l’Ardèche (2018)

2. Akouete, A.F.: Présentation du projet OSM. In: Conférence régionale understanding risk, Abidjan, 20 au 22 Novembre 2019
3. Jansen, M., Mayer, C.: Towards GeoExt 3, Supporting both OpenLayers 3 and ExtJS6, forumFOSS4G 2015, Seoul, South Korea, 18 September 2015
4. Gratier, T., Spencer, P., Hazzard, E.: OpenLayers 3: Beginner's Guide. Packt Publishing, Birmingham (2015)
5. Devillers, R., Jeansoulin, R.: Fundamentals of Spatial Data Quality, p. 312. ISTE Publishing Company, London (2006)
6. Steiniger, S., Hunter, A.J.S.: Free and Open Source GIS Software for Building a Spatial Data Infrastructure, pp. 247–261 (2011)
7. http://www.opengeospatial.org/ consulté le, 15 June 2016

Construction Method of City-Level Geographic Knowledge Graph Based on Geographic Entity

Junwei Liu(✉), Dahai Guo, Guokun Liu, Yanli Zhao, Wenxue Yang, and Liping Tang

Terra InfoTech (Beijing) Co., Ltd., 5F, Block A&B, Ling Zhi Center, Aobei Technology Park, Haidian District, Beijing, China
liujunwei@terra-it.cn, gl1870@nyu.edu

Abstract. Geographic knowledge is a human's summary of the characteristics and laws of geographic things. In recent years, geographic knowledge graph has gradually become a research hotspot. The construction of geographic knowledge graphs uses a large number of general techniques and methods in the field of knowledge graphs, but geographic knowledge graphs also have characteristics that are different from knowledge graphs in other fields. Based on the general construction technology of knowledge graphs and the special nature of geographic knowledge, combined with actual application scenarios and actual development experience, this paper proposes a city-level static geographic knowledge graph construction method based on a single source of surveying and mapping data. The research results of this article effectively solve the problem of using a single source of information to construct the basic framework of the city-level geographic knowledge graph, and lay the foundation for the subsequent integration, enrichment and application of the geographic knowledge graph. At the same time, it has a certain degree in urban management, urban planning and other fields. Versatility has important reference value for the application of geographic knowledge.

Keywords: Geographic entity · Geographic knowledge representation · Geographic knowledge formalization · Geographic knowledge graph

1 Introduction

Knowledge graph is a form of graph organization that associates various entities through semantic association, which is an extension of semantic network technology. The semantic network organizes different data into the form of a graph [1]. The nodes in the graph represent objects or concepts, and the edges represent the relationships between objects or concepts. Semantic network is the intersection of computer information storage technology (database technology) and human conceptual thinking system. By storing data in a form closer to human natural thinking mode, semantic network enables computers to perform human-like reasoning and thinking.

On the basis of the semantic network, the knowledge graph standardizes the semantic model standards of nodes and edges in the semantic network, and introduces statistical analysis of data, making the knowledge graph not only in the traditional intelligent

S. Bourennane and P. Kubicek (Eds.): ICGDA 2022, LNDECT 143, pp. 133–142, 2022.
https://doi.org/10.1007/978-3-031-08017-3_12

systems based on rules and logic (such as knowledge engineering and Expert system) has played a key role, and has excellent applications in natural language processing, search engines, data mining, intelligent question answering and other fields. The knowledge graph extracts and fuses structured and unstructured data together through data, embodies the ideas of data governance and semantic connection, and is conducive to the utilization and migration of large-scale data.

Geographic knowledge is a human's summary of the characteristics and laws of geographic things. As the foundation of human existence and life, the geographic environment directly affects the economic activities, political decision-making and cultural life of human society [2]. The current geographic knowledge information field is characterized by a large number of pure geographic data lacking semantics. These data are obtained by means of surveying and mapping, but without a conceptual network incorporating geographic knowledge, it is difficult to directly contribute to urban management and urban decision-making. As a result, geographic knowledge graph has gradually become a research hotspot in recent years.

The construction process of geographic knowledge graph is usually divided into four steps: the first step, data extraction: select a certain geographic instance in the relational database to construct the ontology database. The second step, semantic extraction and association: extract instance attributes, use identification codes and geographic entity spatial identity codes as indexes, associate the instances with corresponding geographic entities in the basic geographic entity database, and realize the semantic expansion of geographic unit instances. The third step, semantic description: build a semantic relationship table based on the semantic information of the extended domain. The fourth step, map construction: After knowledge verification and other processing procedures, semantic information is converted into domain knowledge and expressed in graphs.

Geographic knowledge graphs use a large number of general techniques and methods in the field of knowledge graphs [3], but geographic knowledge graphs also have characteristics that are different from those of other domains: (1) Accuracy. Geographic information is of great importance to human production practices. Therefore, compared with knowledge graphs such as social network knowledge graphs or e-commerce information knowledge graphs, the data of geographic knowledge graphs requires higher accuracy and reliability. This means that, When performing information extraction and knowledge fusion, the source of information needs to be more carefully identified. (2) Relatively static. The geographic knowledge graph should be used as the basis for the application of knowledge in other social fields. The update and modification of the geographic knowledge graph will involve many other fields, so the core data should remain relatively static. (3) Semantic scarcity. The geographic data obtained from surveying and mapping usually only has classification, parameters and geometric models, and cannot be directly transformed into concepts and objects with natural semantics.

An important function of the knowledge graph is to fuse multi-source heterogeneous data into a unified knowledge system. However, because of the characteristics of the above-mentioned geographic knowledge graph, we often need to use a single source of data to establish a relatively static picture frame. This frame is derived from structured surveying and mapping data and provides a basis for the development and construction of the subsequent knowledge graph. This article will introduce some ideas, methods and possible problems in the construction of the basic framework of the city-level geographic knowledge graph.

2 Geographic Knowledge Graph Structure Design

Ontology design is the first step in the construction of knowledge graph. It provides a formal, explicit specification of a shared conceptualization. The traditional geographic ontology divides geographic entities into two categories: Ecosystem (ocean, forest, desert, etc.) and Landform (channel, bay, mountain, etc.). This classification is semantically correct, but not suitable for knowledge graph construction. It is more common today to classify geographic entities into three categories, areas, points, and lines, based on their geometric models [4]. This means that the spatial attributes of geographic entities rather than semantic attributes are used as the basic framework for the ontology design of geographic knowledge graphs.

Generally speaking, the attributes of a complete geographic entity should include the following six parts: space, state, evolution, change, interaction, usage [5]. Among these six parts, space and state focus on the static conditions of objects, evolution and change focus more on the dynamic conditions of objects, and interactions and usage depend on the relationships and mechanisms between geographic objects. Therefore, in the initial construction process of geographic knowledge graph, space and state attributes are the most important.

The knowledge graph combines the functions of computer storage query technology and human conceptual reasoning, so there are two different data storage representation models: RDF triples and graph databases. RDF is more inclined to the expression and reasoning of concept systems. Through the RDF Schema and OWL language based on RDF, objects and concepts are clearly distinguished, and there is a convenient reasoning engine that can be used. The graph database pays more attention to data storage and query, and does not require very elaborate ontology design, and is more suitable for industrial scenarios. The characteristics of the framework construction phase of the city-level geographic knowledge graph are that the amount of data is large and the conceptual framework is relatively simple, so this article adopts the data storage and representation model based on the graph database neo4j.

2.1 Classification of City-Level Geographic Entities

A geographic entity is a geographic object that occupies a certain and continuous spatial position in the real world and has the same attribute or complete function. Geographic entity data is the "abstraction" of geographic entities in a computer environment. It has semantic recognition meaning and geospatial features, and geospatial features are features with geographic spatial location attributes. Geographic entity data can have a variety of expression forms, such as vector 2D data, remote sensing image data, refined 3D model data, oblique photography real scene 3D model data, laser point cloud data, etc.

City-level geographic entities are divided into two categories: geographic units and geographic physical entities. The geographic unit is essentially a two-dimensional plane block, including administrative division unit, land use unit, management unit and physical geographic unit, etc. Geographic object entities take into account both two-dimensional and three-dimensional expressions, including entities such as water systems, traffic, buildings (structures), pipelines, and courtyards (Fig. 1).

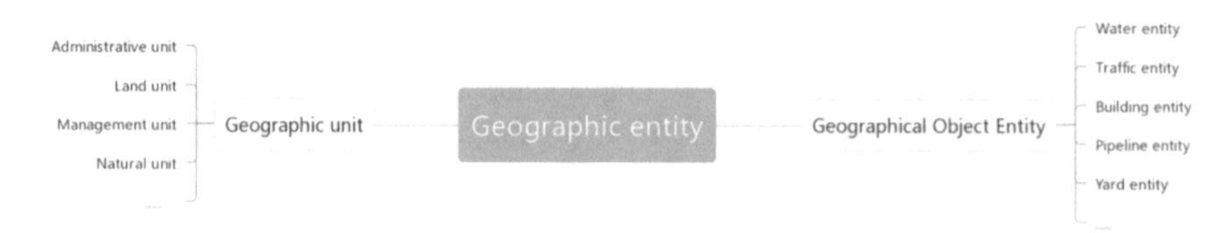

Fig. 1. City-level geographic entity classification

The surveying and mapping data of each geographic entity is divided into three parts: identification information, spatial information, and attribute information.

(1) Identification information refers to the basic identification information of an entity, including name, identity code, location code, address, etc.
(2) Spatial information mainly describes the spatial characteristics of geographic entities, including position, orientation, direction, direction, range, size, height, depth, shape, distribution, etc. When abstracted as geometric shapes, use points, lines, surfaces, bodies, etc. Geometry expression can also be expressed in the form of images, three-dimensional models, oblique photogrammetric models, etc.
(3) Attribute information mainly describes the intrinsic and essential characteristics of geographic entities that distinguish individual differences, including classification, use, length, area, material, periodicity (such as seasonal cycle, tidal cycle), etc.

In the city-level geographic knowledge graph, the geographic entities as the unified spatial positioning framework and the basis of spatial analysis are the basic geographic entities. They are the basis of the entire geographic knowledge graph. All other geographic entities are derived from basic geographic entities and are called derived geographic entities.

There are three types of derived geographic entities: one is combined geographic entities, which are determined by the same management authority, such as courtyard combination entities; the second is aggregate geographic entities, which are determined by the same functional theme, such as road intersections; the third is thematic geographic entities, which are determined by the same Determined by the attribute type, such as an urban green space entity.

2.2 The Hierarchical Relationship of Geographic Knowledge Graph

The nodes in the knowledge graph represent an entity or concept. In order to have a clear structure and good scalability of the graph, a tree graph should be constructed vertically, and the basic geographic entities should be used as the leaf nodes of the entire graph. According to different relationship attributes, two different tree structures can be constructed (Fig. 2).

According to the generic relationship, the categorical relationship tree can be constructed: starting from the basic geographic entity node, the entity-concept tree is constructed in the order of "basic entity-derived entity-concrete concept-abstract concept". Taking the road entity as an example, a branch of the entity-concept tree is similar to "XX road segment 1 (basic geographic entity)-XX road (aggregated geographic entity)-urban road (specific geographic concept)-unclosed road-transportation system (abstract Geographic concept)-physical entities". The generic relationship is divided into three categories according to the difference of semantics: (1) Whole-part relationship, between the derived geographic entity and the basic geographic entity. (2) Concept-instance relationship, between geographic concepts and geographic entities. (3) The species relationship, between abstract geographic concepts and specific geographic concepts.

According to the entity's spatial geographic location relationship, a spatial relationship tree can be constructed: starting from the basic geographic entity node, the lower-level entity should be regarded as a point relative to the upper-level entity in the spatial scope, otherwise it should be treated as the same-level entity. Taking the road entity as an example, a branch of the spatial concept tree is similar to "XX road segment 1 (basic geographic entity)-XX courtyard (range geographic entity)-XX administrative district (geographic unit)-XX city".

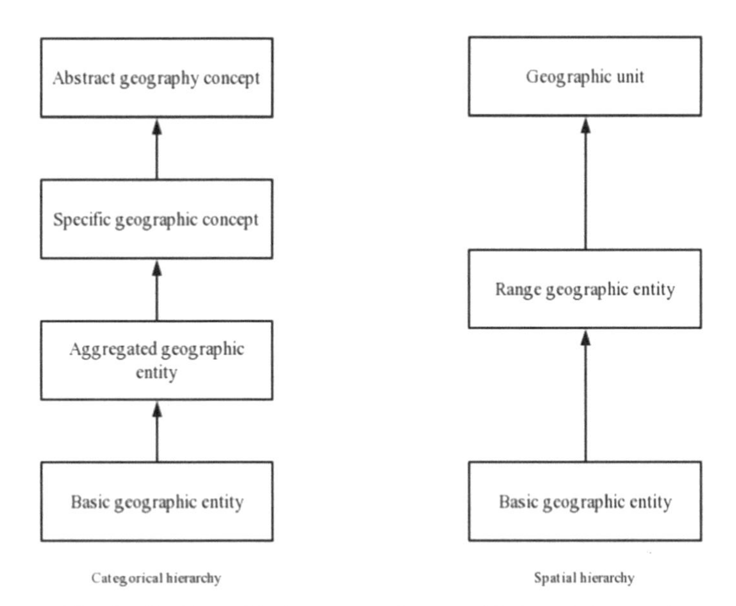

Fig. 2. The hierarchical structure of geographic knowledge graph

In addition to the above two most general geographic knowledge graph hierarchical relationships, it can also be layered based on other existing data. For example, based on administrative management relationships, a management-level relationship tree can be constructed, and so on.

2.3 Fusion of Multi-level Geographic Knowledge Graphs

In practical applications, the granularity of the division of geographic entities is not unique according to different requirements and scenarios, so there may also be a finer structure within the basic geographic entities. For example, in the city-level geographic knowledge graph, the feature entity with a three-dimensional model is usually used as the basic feature entity, and the two-dimensional feature entity and geographic unit are usually treated as derived entities as scope entities. Another example, in the city-level geographic knowledge graph, a building is regarded as a basic and indivisible geographic entity, but in the architectural-level geographic knowledge graph, a building will have "doors and windows-rooms-households-floors". "Such an internal layered structure". Therefore, in the design of the geographic knowledge graph structure of each granularity, an "end node" should be set so that all nodes in the geographic knowledge graph at this level are indirectly attributed to this end node, so that the knowledge graphs of different levels can be parallelized. Constructed, and the completed knowledge graph can be easily integrated (Fig. 3).

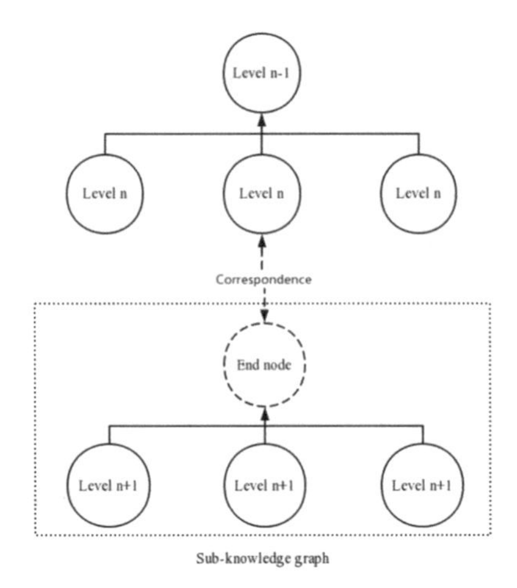

Fig. 3. End node structure

3 The Establishment and Basic Application of the Horizontal Relationship of the Knowledge Graph

After the hierarchical tree trunk of the geographic knowledge graph is established, the horizontal relationship between nodes at the same level should be established. These horizontal relationships can be divided into three categories: belong-together relationships, spatial relationships, and derived relationships (Fig. 4).

3.1 Determine the Belong-Together Relationship Through the Hierarchical Structure

If several geographic entities jointly constitute or belong to an upper-level entity, such as road intersections, traffic lights, overpasses, and underpasses that form a road intersection, then the above entities have a belong-together relationship. The characteristics of belong-together relationships are qualitative but not quantitative. The belong-together relationship can be more directly derived from the categorical relationship, so it does not necessarily need to be explicitly constructed depending on the situation. The scenarios for establishing a belong-together relationship are: (1) When the combination or aggregation relationship of some entities is relatively stable, but the upper-level entities they belong to may change (for example, the buildings, roads, and auxiliary facilities in the courtyard are stable as a functional aggregate, But the name and scope of the corresponding institution may change). (2) When access to the knowledge graph is authorized (for example, a building-level geographic knowledge graph cannot access the information of the upper courtyard node, but can access the information of other building nodes of the same level).

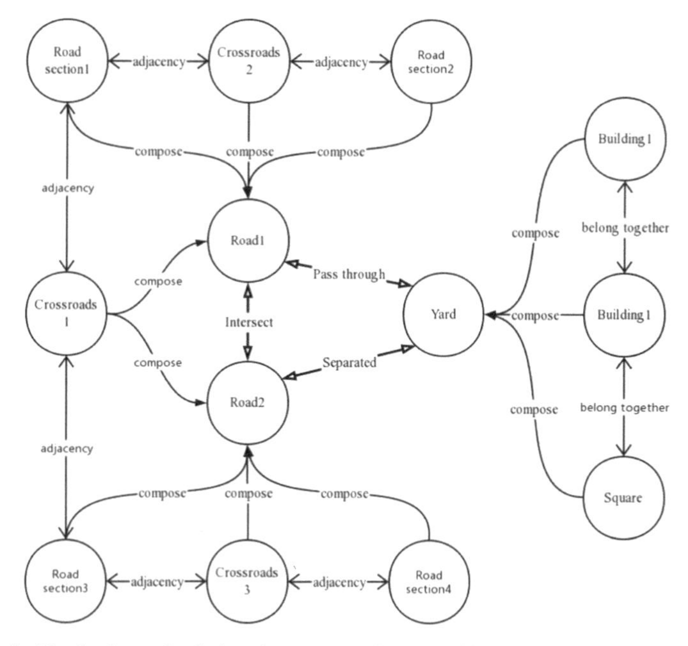

Fig. 4. The horizontal relational structure of geographic knowledge graph (partial)

3.2 Calculate the Spatial Relationship Through the Geometric Model

Spatial relationship is the description of the relationship between the relative spatial positions of geographic entities. According to the quantifiable characteristics of spatial semantic relations, its description can be divided into two types: qualitative relation description and quantitative relation description. According to the quantifiable characteristics of the spatial relationship, it can be divided into spatial topological relationship (basic spatial relationship, qualitative description), spatial distance relationship (discrete or continuous quantitative description) and orientation relationship (qualitative description or quantitative description).

By calculating the geometric model data of the geographic entities, the spatial quantitative relationship between the geographic entities can be obtained, and then the conversion mapping rules between the quantitative description and the qualitative description can be specified to obtain the spatial qualitative relationship between the geographic entities. For the city-level geographic knowledge graph, the most important thing is the calculation of two-dimensional and sub-two-dimensional geometric models. The most common calculations involve: the containment relationship between points and areas, the intersection relationship between lines and lines, the intersection relationship between lines and areas, the intersection relationship between faces and faces, and the containment relationship between faces and faces. For the same geographic entity, when calculating different spatial relationships, different geometric models need to be used, so the geometric models need to be transformed. Common transformations include: calculating the two-dimensional projection of the three-dimensional model, and calculating the boundary line of the courtyard area, calculate the centerline of tubular geographic entities (such as road sections and river sections), and calculate the center point of its two-dimensional projection for smaller geographic entities.

3.3 Calculating Derived Semantic Relations by Deriving Geographic Entities

The spatial relationship calculated by the geometric model is usually the adjacency relationship between basic geographic entities. Combining several spatial relationships can obtain a more abstract relationship between derived entities, which is called derived semantic relationship. Compared with spatial relations, derived semantic relations have less stability, but have richer semantic content, and can carry out enlightening data mining and information analysis. For example, the staggered sequence of road sections and road intersections constitutes a derivative entity of a complete road, and the intersection relationship between roads and roads can be constructed based on the adjacency relationship between road sections and staggered roads. In addition, derived semantic relationships establish upper-level semantic relationships between geographic entities of different types and levels that do not directly involve geographic locations. For example, the power supply relationship between the substation and the building can be constructed through the affiliation relationship between the substation and the transmission line, and the adjacent relationship between the wire and the building. This relationship is independent of the actual spatial location of the transmission line.

3.4 Information Mining Algorithm Based on Graph Relation

The horizontal relationships in the knowledge graph establish graph relations among geographic entities at the same level, so common graph analysis methods can be used to analyze these data. There are three common applications of graph analysis: path search, centrality analysis, and community discovery [6].

The main algorithms of path search include graph traversal, shortest path search, minimum spanning tree, random walk, etc. There are many horizontal networks with real distance as the edge weight in the geographic knowledge graph, so there are important applications in the field of traffic structure analysis.

Centrality analysis measures how a node is the "central node" of the graph according to different indicators. The main indicators include degree centrality, closeness centrality, and betweenness centrality. Commonly used algorithms are PageRank algorithm. Centrality analysis can find the most important node in a graph, and it plays a key role in the flow control and fault warning of transportation network, water system network, electric power and water conservancy network.

The community discovery algorithm calculates the closeness of the connection between the set of nodes in the graph, and finds the set of nodes with the closest connection. Commonly used algorithms include measuring algorithm, components algorithm, label propagation algorithm, etc. The community found that it has an auxiliary effect on the decision-making of some social policies, such as the division of management areas, the allocation of public resources, and the choice of transportation hub locations.

It should be noted that the graph analysis algorithm based on a single-source static geographic knowledge graph is not accurate, and the effectiveness of the specific algorithm needs to be adjusted according to more detailed and accurate data.

4 Conclusion and the Following Research Directions

Based on the general construction technology of knowledge graphs and the special nature of geographic knowledge, combined with actual application scenarios and actual development experience, this paper proposes a city-level static geographic knowledge graph construction method based on a single source of surveying and mapping data. The main process has two steps: first, determine the vertical hierarchical tree structure of the geographic knowledge graph according to the classification of city-level geographic entities, and then determine the horizontal network structure between each level of the geographic knowledge graph according to the semantics of the geographic entities. In addition, some graph analysis algorithms that can operate on horizontal structures are also proposed.

The following research directions mainly include three areas: One is to construct asymmetrical horizontal relationships. The spatial relationships (adjacent, intersecting) obtained based on geometric model calculations are usually symmetrical, and simple geometric calculations are difficult to establish between geographic entities. Sequential relationship, such as the upstream and downstream relationship of a river section or the front and back relationship of a one-way road section, requires more complex geometric calculations and combined with other attributes of the entity. The second is to make full use of the three-dimensional model of geographic entities. For geographic entities such as overpasses, tunnels, and bridges, two-dimensional data is far from enough to characterize their attributes. The third is to clarify the priority and specific methods of other unstructured multi-source data when modifying and expanding the static framework of the geographic knowledge graph.

References

1. Fensel, D., et al.: Introduction: what is a knowledge graph? In: Knowledge Graphs. Springer, Cham (2020). https://doi.org/10.1007/978-3-030-37439-6_1
2. Wei, Z., Wan, G., Mu, Y., et al.: Design and construction of geographic knowledge graph. In: 2020 IEEE 9th Joint International Information Technology and Artificial Intelligence Conference (ITAIC), vol. 9, pp. 2252–2256 (2020)
3. LiuQiao, L.Y., DuanHong, L.Y.: Knowledge graph construction techniques. J. Comput. Res. Dev. **53**(3), 582 (2016)
4. Laurini, R., Kazar, O.: Geographic ontologies: survey and challenges. Meta-carto-semiotics **9**(1), 1–13 (2017)
5. Wang, S., Zhang, X., Ye, P., Du, M., Lu, Y., Xue, H.: Geographic knowledge graph (GeoKG): a formalized geographic knowledge representation. ISPRS Int. J. Geo-Inf. **8**, 184 (2019). https://doi.org/10.3390/ijgi8040184
6. Needham, M., Hodler, A.E.: Graph Algorithms: Practical Examples in Apache Spark and Neo4j. O'Reilly Media (2019)

Development of Remote Sensing Software Using a Boosted Tree Machine Learning Model Architecture for Professional and Citizen Science Applications

Suraj N. Vaddi(✉) and Kathleen M. Morrow

Thomas Jefferson High School for Science and Technology, Alexandria, VA 22312, USA
surajnvaddi@gmail.com

Abstract. Remote sensing has rapidly gained significance in environmental science research over the past twenty years, especially when it comes to assessing the urbanizing landscape. As urban sprawl continues to grow, the role of citizen science in environmental studies has also gained greater importance by providing a broader range of study routes. The current study aims to bridge the gap between remote sensing of urbanization and the ability of citizen scientists to contribute to our expanding knowledge base. We developed a machine learning model-based architecture that makes use of pre-trained models to estimate pervious and impervious surface percentages within a user-defined region. Pre-trained model-based architectures provide greater ease-of-use and can be made more accessible than softwares that require manual supervised learning (e.g. ArcGIS). Therefore, we have developed an iOS application called Tar Print that utilizes model-based architecture, with a boosted tree algorithm and per-pixel classification of publicly available satellite imagery. We optimize the iterative per-pixel design by extracting features about neighboring and contrasting pixels, which are conventionally under-utilized, to provide the machine learning framework with more features. The Tar Print version 1 model includes six training classes and nine features that are sorted into both pervious and impervious categories. We used 2,250 data points to train this model, which achieved a 100% training score, 98% testing score, and an 86% validation score. We are currently developing a version 2 model that is trained over 50,000 data points and nine training classes. This research expands the tools citizen scientists and professionals can use to accurately monitor urban development on a larger scale.

Keywords: Remote sensing · Geoinformatics · Impervious surfaces · Urbanization · Citizen science · Boosted tree · Model-based architecture · Supervised learning · Machine learning

S. Bourennane and P. Kubicek (Eds.): ICGDA 2022, LNDECT 143, pp. 143–152, 2022.
https://doi.org/10.1007/978-3-031-08017-3_13

1 Introduction

Remote sensing has rapidly gained traction in the field of environmental science over the past twenty years [1], leading to a large body of supporting literature and the development of remote sensing applications. Traditionally, remote sensing uses artificial intelligence (AI) and machine learning (ML) techniques to extract data from various types of satellite imagery. These AI and ML techniques include supervised learning and unsupervised learning [2].

Integrating citizen science with ecological research has quickly gained popularity. Usually, AI and ML tools support citizen science research by finding and predicting trends within large datasets. Citizen science allows researchers to account for a larger range of variables and to reach a broader set of data sources, tracking changes that would be impractical to study on a smaller scale [3].

One such field is urban development and the study of impervious surface cover, which includes roads, buildings, and sidewalks. Impervious surfaces prevent water from entering the ground [4] and are highly associated with urbanized settings. The lack of permeability prevents water from recharging the ground water supply but increases discharge during periods of high rainfall into streams, which leads to flooding and stream bank erosion. This variability in the watershed leads to instability of stream banks, water quality and, overall stream health [5, 6]. Impervious surfaces also excessively heat water [7] and collect chemicals and pollutants including road salts [8] and gardening chemicals [9]. These factors reduce water quality and increase the economic burden on stormwater treatment and water purification facilities [10].

As migration into urban areas increases exponentially [11], the negative impacts of urbanization and impervious surfaces are becoming more paramount to study and monitor on a large scale, making it a practical candidate for citizen science efforts. In fact, a number of methods have already been developed to analyze impervious surface cover. For example, neural networks are often used to supply a machine learning model with the raw unclassified data so that the model itself can create connections and patterns among training features. Commonly used neural networks include convolutional, recurrent, and deep neural networks that are trained using temporal or spatial information (in the form of convolutions) that provide the machine learning framework with context to data points [12–15]. Additional machine learning frameworks require raw data to be classified before the model can be trained. For example, decision trees use recursive partitioning to extract features, and random forests employ multiple decision trees to achieve more accurate results. Gradient boosted decision trees are more powerful than both as they use multiple weak decision trees but use gradient boosting to make progressively improved predictions [16]. However, as machine learning models become more accurate, they also require more training points and more specific features and classes.

Current remote sensing applications to quantify pervious and impervious surface types make use of machine learning models. However, they can often be challenging to use because they require users to manually train data sets of various surface types. This limits the extent to which citizen science efforts can aid in monitoring urbanization. To fill this need, we developed Tar Print, a remote sensing software application that is specifically designed to monitor impervious surfaces in urban settings. Our goal is to enhance citizen science efforts with Tar Print's straightforward design that uses a model-based machine learning approach and visible satellite imagery embedded within a user-friendly iOS application. We combine convolutional techniques along with machine learning models that utilize pre-classified data sets from Apple Maps satellite images, using the gradient boosted decision tree architecture. The current results analyze the machine learning model of Tar Print version 1 (compatible with iOS 14) which we released to the App Store in December 2021. We also provide a discussion of the prospective improvements we are implementing with Tar Print version 2, which is yet to be published and will be compatible with iOS 15 [17].

2 Design and Algorithms

2.1 User Interface

A primary goal while developing Tar Print was to address barriers of entry into remote sensing research by increasing Tar Print's ease-of-use. We do this by integrating Apple Maps satellite imagery from MapKit [18] which is a familiar interface for citizen scientists and researchers alike and makes testing in the field more accessible. A user can analyze the surface type around a location using their current location, placing a pin on a location, or by searching for a location with latitude and longitude coordinates. Integration with Apple Maps streamlines the process of retrieving coordinate points. Furthermore, the user has the option to select the side length of the satellite image as 1.0, 2.0, or 3.0 miles. These are the only steps a user must take before processing their coordinate point (Fig. 1).

The processing methods combine machine learning and convolutional algorithms that extract details about each pixel on the image to determine if the pixel is pervious or impervious (Fig. 1). Once image processing is finalized, the satellite image of the region is displayed along with an overlayed image that displays the permeability of the image. Each pixel on the permeability image is colored based on the surface type, and below it the percentages of impervious surface, pervious surface, and water are shown.

2.2 Image Processing

We used MapKit [18] functionality to interact with and capture satellite imagery, and we used CreateML [19] to train and evaluate the machine learning model. When the user defines a location to quantify based on latitude and longitude coordinates, MapKit captures a satellite image (200 by 200 pixels) of the surrounding region, and the image processing algorithm operates on a per-pixel basis to categorize each pixel as a variety of impervious or pervious surfaces. The surface types include dry pervious, shaded pervious, fall pervious, tan pervious, standard pervious, and standard impervious.

Using a machine learning model to classify each pixel is time-intensive, so we use a series of filters to reduce the number of pixels that require classification by the machine learning model, namely bodies of water and pervious surfaces that can be clearly defined by bright green hues (Fig. 1).

Bodies of water are filtered by cross-analyzing satellite image pixels with Apple Map's overlayed imagery. Pixels that are clearly green are filtered using an inequality of red (r), green (g), and blue (b) values compared with thresholds to determine if the pixel is pervious surface (Eqs. 1 and 2). Satisfying either inequality filters the pixel as pervious.

$$36 \leq 2.03g - r - b \tag{1}$$

$$g > r + 2b - 75 \tag{2}$$

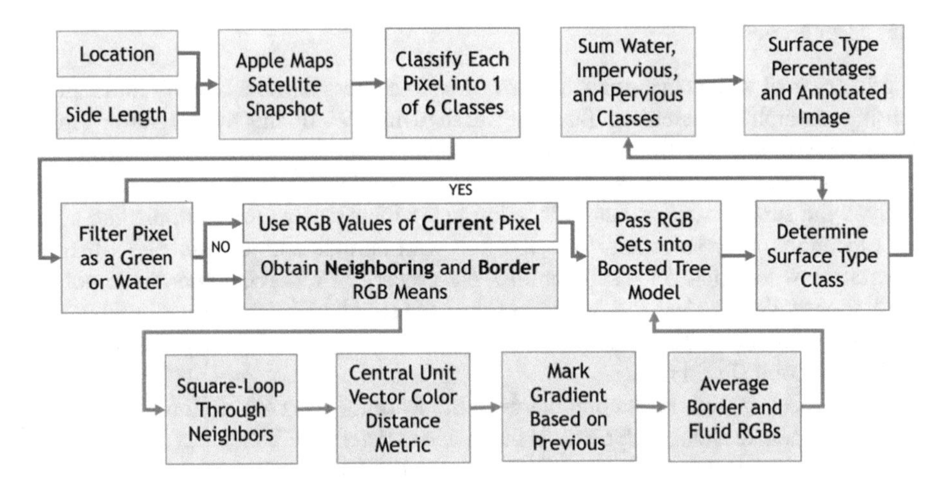

Fig. 1. Flowchart of algorithm used by Tar Print in analyzing a plot of satellite imagery. The model receives coordinates (location) and the side length of the square plot to analyze to report percentages of impervious surface, pervious surface, and water surface types. The first row outlines the high-level steps the algorithm takes, the second row outlines the machine learning model process, and the bottom row outlines the steps for feature extraction that optimize this model.

2.3 Feature Extraction and Convolutions

The feature extraction process (Fig. 1) allows for the machine learning model to make more accurate predictions based on a larger number of predictors. Due to the noise gained by reducing a miles-long plot of satellite imagery to a 200 by 200-pixel grid, extracting features about neighboring pixels is essential for producing more accurate results.

If a pixel is not filtered, Tar Print further assesses that pixel using a convolution to extract more features about its surroundings. These include the red, green, blue (RGB) value of the pixel, the average RGB value of surrounding, similarly colored pixels (local average), and the average RGB value of a differentiating edge strip (edge average). Thus, nine features are computed for each pixel: red, green, blue, local red, local green, local blue, edge red, edge green, and edge blue.

The size of the convolution depends on the area of the plot. Each pixel on larger satellite images represents a greater area than pixels on images with shorter side lengths. Thus, pixels on images of greater surface area require fewer surrounding pixels to be characterized. Therefore, the width of the convolution applied to each pixel (w) decreases as the user-determined satellite image side length increases (l) to measure only essential pixels that characterize the surrounding region (Eq. 3).

$$w = -5l + 28 \tag{3}$$

After the width of the convolution is determined, we apply the feature extraction algorithm (Fig. 1). In essence, this algorithm analyzes the gradient of color change between concentric squares to determine where there is a large enough color difference to constitute a border, and these differences are stored in an edge matrix. The feature extraction algorithm operates in two stages. The first stage differentiates pixels that are local, represented as a 0 in the edge matrix, and those that are edges or follow edges, represented as a 1 in the edge matrix (Fig. 3A). The second stage determines the strip of pixels that constitutes the immediate edge within the convolution (Fig. 3B).

At each iteration of stage one, pixels on the border of the current square are compared to pixels on the border of the previous iteration. The direction in which current pixels must check for their corresponding pixels on the previous iteration is assessed using the unit vector the points to the center of the image staggered by 45°. Through this method, the vertices of the current square point to the vertices of the previous square iteration.

For each comparison there are three possible outcomes: (1) the previous pixel was an edge, (2) the previous pixel was not an edge and there is no color difference between the pixels, and (3) the previous pixel was not an edge and there is a color difference between the pixels.

Commonly in edge detection, pixels are compared using their intensities, the averages of their red, green, and blue values, known as the intensity. However, this method does not account for pixels that have similar intensities but different ratios of red, green, and blue. Impervious surface mapping often deals with grey colors, so the intensity of differently colored pixels may appear to be similar. Therefore, rather than comparing intensities, we calculate a concentrated RGB difference that measures the absolute difference in red, green, and blue values of two pixels (Eq. 4). The differences between outer and inner RGB values are given by Δr, Δg, and Δb. The pixel on the current square is labeled as a border if the concentrated absolute RGB difference exceeds 15. We call this method: concentrated intensity difference (Fig. 2).

$$\frac{|\Delta r| + |\Delta g| + |\Delta b|}{3} \geq 15 \tag{4}$$

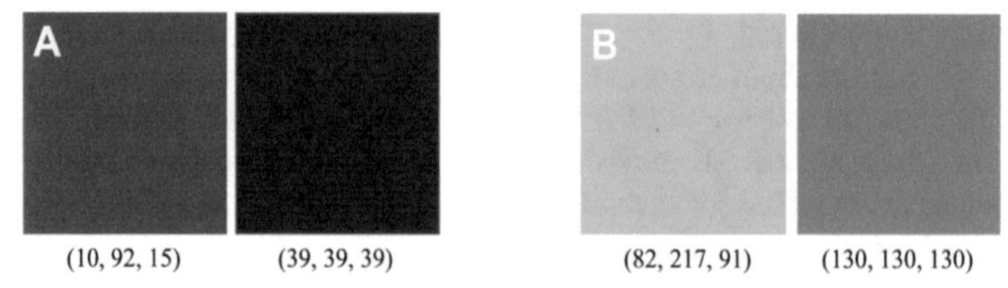

Fig. 2. We use two-pixel pair scenarios with RGB values beneath each pixel to demonstrate two cases in which intensity difference does not detect a color difference, but concentrated intensity difference does. In pixel pair A, the intensity of each pixel is 39, so no color difference is detected using intensity difference. However, the concentrated absolute RGB difference value (Equation) is 35 – detecting the color difference. In pixel pair B, the intensity of each pixel is 130, so no difference is detected once again. However, the concentrated absolute RGB difference value is 58 – again detecting the color difference.

After stage one, the generated edge matrix is used to determine immediate border pixels for edge average calculation (Fig. 3B). We call this ensemble of outward edge detection and hysteresis thickening: gradient source edge detection.

A

1	1	1	1	1	1	1	1	1	1	1
1	1	1	1	1	1	1	1	1	1	1
1	1	0	0	0	1	1	1	1	1	1
1	1	0	0	0	0	0	0	1	1	1
1	1	0	0	0	0	0	0	0	1	1
1	1	0	0	0	x_0	0	0	0	1	1
1	1	0	0	0	0	0	0	0	1	1
1	1	0	0	0	0	0	0	0	1	1
1	1	0	0	0	0	0	0	0	1	1
1	1	1	1	1	1	1	1	1	1	1
1	1	1	1	1	1	1	1	1	1	1

B

3	3	3	3	3	4	4	4	4	4	4
3	2	2	2	2	3	3	3	3	3	4
3	2	0	0	0	2	2	2	2	3	3
3	2	0	0	0	0	0	0	2	3	3
3	2	0	0	0	0	0	0	0	2	3
3	2	0	0	0	x_0	0	0	0	2	3
3	2	0	0	0	0	0	0	0	2	3
3	2	0	0	0	0	0	0	0	2	3
3	2	0	0	0	0	0	0	0	2	3
3	2	2	2	2	2	2	2	2	2	3
3	3	3	3	3	3	3	3	3	3	3

Fig. 3. Stages one (A) and two (B) of feature extraction algorithm: gradient source edge detection. For larger convolutions than depicted above, stage two would have pixels labeled 4 surrounding those that are labeled 3, and pixels beyond 4 would remain labeled 1.

2.4 Boosted Tree Architecture

We used a boosted tree algorithm (gradient boosted decision tree) to train our machine learning model using CreateML [19] using the specifications outlined in Table 1. The training dataset contained 2,250 points with each point having nine features: red, green, blue, local red, local green, local blue, edge red, edge green, and edge blue. These points were also associated with a class to label surface type at each pixel. We also achieved optimal testing results as outlined in Table 2 and overall scores of 100% training score, 98% testing, and an 86% validation scores.

The purpose of subdividing pixels into six classes is to allow the machine learning model to allow for more specialized predictions to maximize false positives. For example, a pervious pixel that is considered "shaded pervious" may be classified as "normal pervious," but this incorrect classification does not impact accuracy because both labels are ultimately counted as pervious surfaces. Furthermore, we included the "fall pervious" training class because publicly available satellite imagery has a wide array of foliage, including trees that change color during fall and winter. We also measure both "dry pervious" and "tan pervious" class that are ultimately combined, so while we trained the machine learning model using six classes, only five are measured using the machine learning model: tan pervious, fall pervious, shaded pervious, pervious, and impervious. However, water is also a measured class that only appears in the filtration stage.

Table 1. Boosted tree model specifications.

Attribute	Metric
Max Iterations	30
Max Depth	25
Min Loss Reduction	0
Min Child Weight	0.05
Row Subsample Ratio	0.6
Column Subsample Ratio	0.6
Step Size	0.299

Table 2. Training scores for machine learning model of Tar Print version 1.

Class	Training		Validation		Testing	
	Precision	Recall	Precision	Recall	Precision	Recall
Dry Pervious	100%	100%	89%	67%	100%	100%
Fall Pervious	100%	100%	71%	94%	100%	100%
Shaded Pervious	100%	99%	88%	54%	97%	97%
Tan Pervious	100%	100%	100%	75%	100%	100%
Pervious	100%	100%	83%	80%	100%	96%
Impervious	100%	100%	93%	100%	92%	100%

2.5 Surface Type Quantification

Every pixel on the satellite plot being analyzed undergoes filtration, both stages of feature extraction, and analysis by the boosted tree machine learning model. After all pixels are analyzed, the total counts of each class are assessed to form the pervious surface percentage (sum of dry pervious, fall pervious, shaded pervious, tan pervious, and pervious), impervious surface percentage, and water surface percentage (computed during Apple Maps filtration). Overlayed permeability images are also displayed reflecting each pixel of pervious, impervious, or water surface type (Fig. 4).

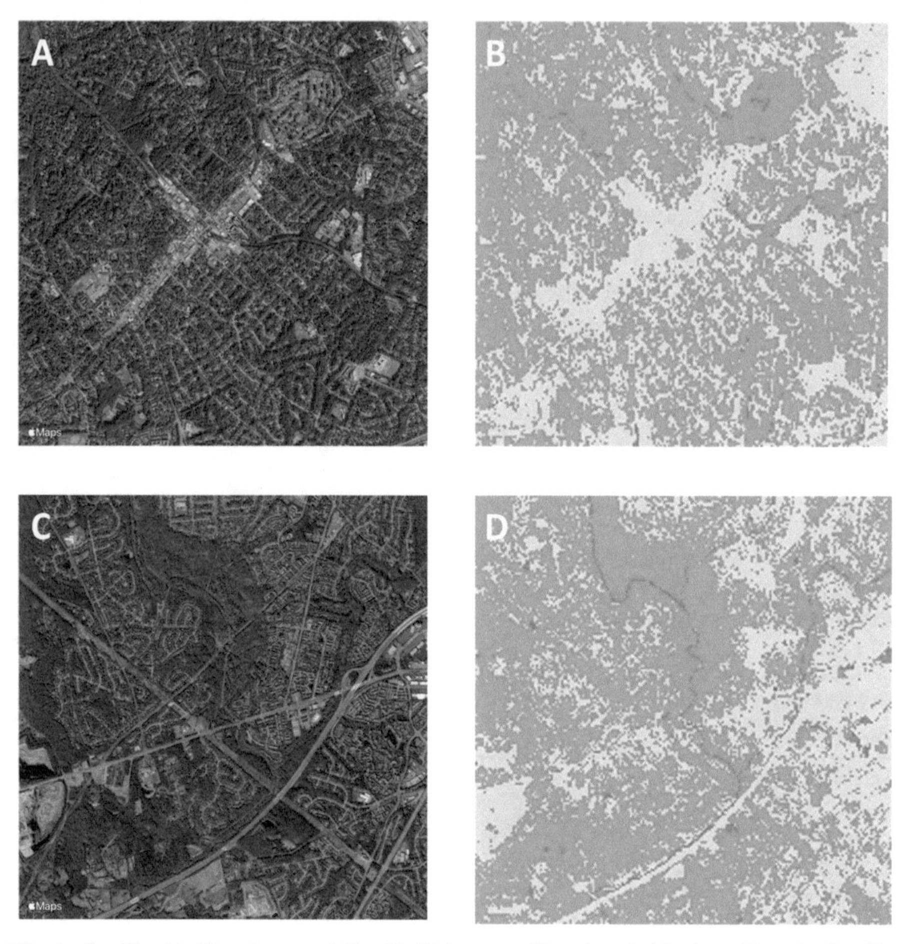

Fig. 4. Satellite (A, C) and permeability (B, D) images of locations in Northern Virginia. Satellite images provided by Apple Maps [20].

3 Conclusions and Future Directions

The goal of Tar Print is to serve as an accurate and accessible impervious surface quantifier to enhance citizen science efforts in monitoring the rapidly urbanizing landscape.

Ultimately, Tar Print operates on three major stages of image processing: filtration, feature extraction, and machine learning model classification. We use new techniques such as the concentrated absolute RGB difference used in edge detection and the feature extraction algorithm ensemble that iterates using concentric squares and centrally directed unit vectors for gradient comparison. These methods are called "concentrated intensity difference" and "gradient source edge detection" respectively. During image processing, pixels are classified into six categories: tan pervious, fall pervious, shaded pervious, pervious, impervious, and water. After all pixels are classified, they are sorted into pervious, impervious, or water surface types. Our model has shown high accuracy with all testing and training scores above 90% and overall scores of 100% training score, 98% testing, and an 86% validation scores.

Future versions of the application (version 2 and beyond) will assess different architectures including random forests and support vector machines [19]. Furthermore, upcoming machine learning models will be trained using architectures that are more accurate by increasing the number of features to include ten classes that are further characterized by hue: brown-green pervious, dark-green pervious, shaded pervious, blue-green pervious, fall pervious, light-tan pervious, dark-tan pervious, pervious, impervious, and water. Water and pervious will continue to be classified during the filtration process. Upcoming models will also be trained using larger datasets of 50,000 data points, in comparison to the current model that uses 2,250 data points.

References

1. Weng, Q.: Remote sensing of impervious surfaces in the urban areas: Requirements, methods, and trends. Remote Sens. Environ. **117**, 34–49 (2012)
2. Lotfian, M.: The partnership of citizen science and machine learning: benefits, risks, and future challenges for engagement, data collection, and data quality. Sustainability **13**(14) (2021)
3. McKinley, D.: Citizen science can improve conservation science, natural resource management, and environmental protection. Biol. Cons. **208**, 15–28 (2017)
4. Kaur, L.: Impervious surfaces an indicator of hydrological changes in urban watershed: a review. Open Access J. Environ. Soil Sci. **4**(1), 469–473 (2019)
5. Booth, D.: Global perspectives on the urban stream syndrome. Freshw. Sci. **35**(1), 412–420 (2015)
6. Arnold Jr., C.: Impervious surface coverage: the emergence of a key environmental indicator. J. Am. Plan. Assoc. **62**(2), 243–258 (1996)
7. Somers, K.: Streams in the urban heat island: spatial and temporal variability in temperature. Freshw. Science **32**(1), 309–326 (2013)
8. Corsi, S.: A fresh look at road salt: aquatic toxicity and water-quality impacts on local, regional, and national scales. Environ. Sci. Technol. **44**(19), 7376–7382 (2010)
9. Phillips, P.: Pesticides in surface water runoff in south-eastern New York State, USA: seasonal and stormflow effects on concentrations. Pest Manag. Sci. **60**(6), 531–543 (2004)

10. Gaffield, S.: Public health effects of inadequately managed stormwater runoff. Am. J. Public Health **93**(9), 1527–1533 (2003)
11. Lombard, H.: Population growth in Virginia slowest in a century as out-migration continues. University of Virginia Weldon Cooper Center for Public Service (2020). https://statchatva.org/2020/01/27/population-growth-in-virginia-slowest-in-a-century-as-out-migration-continues/. Accessed 12 May 2021
12. Huang, F.: Automatic extraction of urban impervious surfaces based on deep learning and multi-source remote sensing data. J. Vis. Commun. Image Represent. **60**, 16–27 (2019)
13. Parekh, J.: Automatic detection of impervious surfaces from remotely sensed data using deep learning. Remote Sens. **13** (2021)
14. Cui, W.: Application of a hybrid model based on a convolutional auto-encoder and convolutional neural network in object-oriented remote sensing classification. Algorithms **11**(1), 9 (2018)
15. Mazzia, V.: Improvement in land cover and crop classification based on temporal features learning from sentinel-2 data using recurrent-convolutional neural network (R-CNN). Appl. Sci. **10**(1), 238 (2019)
16. Ebrahimi, M.: Comprehensive analysis of machine learning models for prediction of subclinical mastitis: deep learning and gradient-boosted trees outperform other models. Comput. Biol. Med. **114** (2019)
17. Vaddi, S.: Tar Print (Version 1.1) [Mobile app] (2020)
18. Apple Developer: MapKit (2021)
19. Apple Developer: CreateML (2021)
20. Apple: Apple Maps [Location in Fairfax County, Virginia] (2021)

Fast Construction of Vegetation Polygons Based on Object-Oriented Method

Fengmin Wu[1(✉)], Xing Liang[1], Jing Yu[1,2], Zhipeng Zheng[1], and Junjun Liang[1]

[1] Chongqing Geomatics and Remote Sensing Center, Chongqing, China
wufengmin@dl023.net
[2] Chongqing University, Chongqing, China

Abstract. Chongqing is a typically mountainous city and its ground cover seems complex and fragmentized. Traditional construction of vegetation polygons based on Fundamental Geographic Information costs large human operation, which seriously affects the statistical analysis and application of geographic information data. This study combined high-resolution remote sensing image and Digital Line Graphic data to construct vegetation polygons based on the object-oriented multi-resolution segmentation method. The method made full use of the spectral characteristics, geometric features and texture features of vegetation information. The study area was selected from a small village of Changshou District in Chongqing, China. In order to construct vegetation polygons, the identification points of vegetation were applied and other technologies such as normalization of vegetation boundary, topology error correction was also applied to improve accuracy of vegetation boundaries lately. At last, the construction of vegetation polygons was validated by manual visual interpretation with field investigation. The result showed, although the accuracy of automated construction of vegetation polygons was 81.5%, efficiency was 8 times more than manual processing.

Keywords: Fast construction · Vegetation · Object-oriented method

1 Introduction

At present, the vegetation datasets were surveying as line data but not area data in most Fundamental Surveying and Mapping field [1]. The vegetation information was expressed by combination of land cover classification lines and vegetation points. The complex boundaries of land cover classification caused large work to construct vegetation polygons using manual method which took disadvantage of batch production. Obviously, the automatic construction and assignment of vegetation polygon could not only adjust to vegetation polygon production, but also improve the production efficiency. Object-oriented classification technology could analyze the spectral, texture, shape and semantic information of object which had high precision over traditional classification method [2–8].

S. Bourennane and P. Kubicek (Eds.): ICGDA 2022, LNDECT 143, pp. 153–161, 2022.
https://doi.org/10.1007/978-3-031-08017-3_14

This paper firstly applied pre-processed Digital Line Graphic data by MicroStation as input parameters of multiresolution segmentation and constructed interpretation rules for vegetation based on eCognition Developer, then assigned detailed vegetation attributes of points to vegetation segmentation results automatically. We smoothed the vegetation boundaries and merged porphyroclastic polygons using programming technology of ArcGIS engine as vegetation polygons had many jagged edges and bumps.

2 Method

2.1 Data

The study area, Bake Town was located in the middle of Changshou District in Chongqing, China (Fig. 1). The Central latitude and longitude was 29°55′48″N, 107°54″E. The image used in the research was worldview-2 in July 2017, including 4 bands (red, green, blue and near-infrared) with high resolution of 0.6 m. The plotting scale of Digital Line Graphic data was 1/5000 which contained vegetation and soil texture points and land cover classification lines. From the Specifications for feature classification and codes of fundamental geographic information, vegetation points were divided into 9 classes in study area (Table 1).

Table 1. Type of vegetation points in study area.

Serial number	Vegetation type
1	Farmland
2	Dry farmland
3	Woodland
4	Arbor forest
5	Shrubbery
6	Bamboo forest
7	Nursery
8	Grassland
9	Half weeds

In this study, different segmentation scales were applied to differentiate the surface types. Then the jagged boundary of split raster data was smoothed by setting distance threshold from dividing boundaries. The vegetation types were identified by attributes of points from Digital Line Graphic data. Finally, the results were validated by field investigation and manual visual interpretation (Fig. 2).

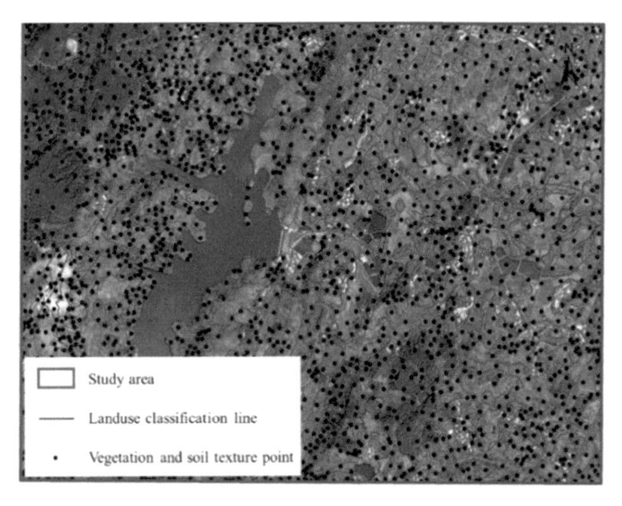

Fig. 1. The study area image and Digital Line Graphic data.

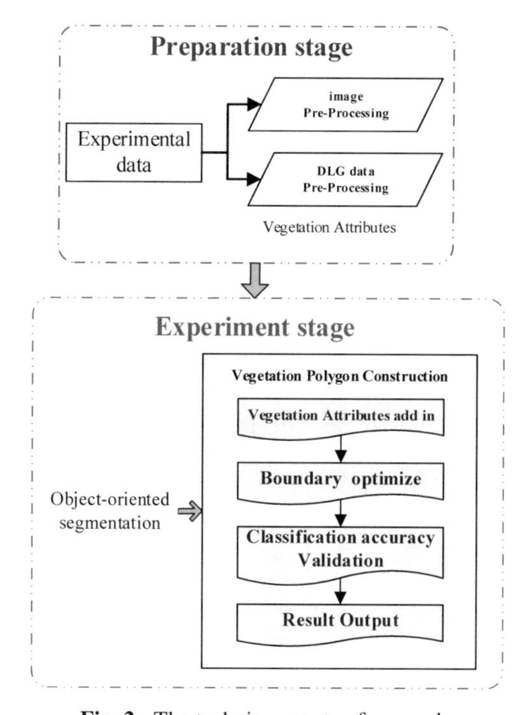

Fig. 2. The technique route of research.

2.2 Multiresolution Segmentation and Classification

Multiresolution segmentation of eCognition Developer was a fundamental and critical step of classification which widely used to identify land use types [9–11]. The parameters of segmentation mainly included spectral factor, shape factor, band weight and segmentation scale. We set the segmentation scales and other parameters according to many times of tests (Table 2). It was very clear that large scale of segmentation seemed more suitable for water area and arbor forest, as small scale was convenient for dispersive buildings and narrow roads with small areas (Fig. 3).

Table 2. Parameters of multiresolution segmentation.

Image layer	Level1	Level2
Segmentation scales	100	50
Band factor	1, 1, 1, 1	1, 1, 1, 1
Shape factor	0.2	0.3
Compactness factor	0.5	0.6

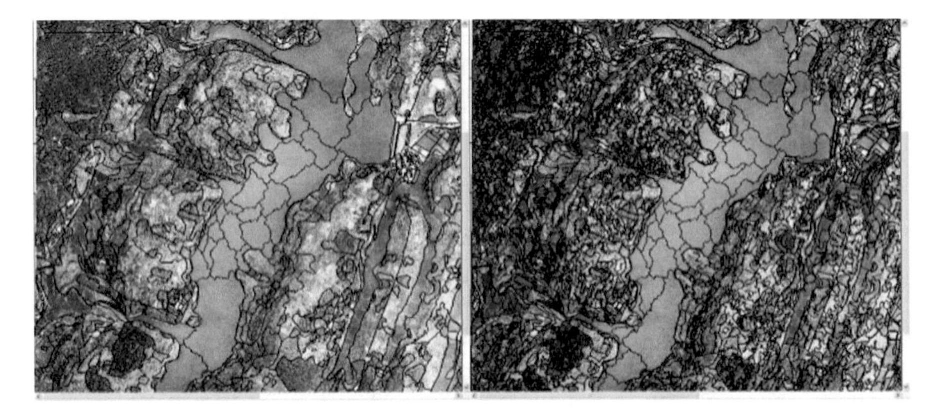

Fig. 3. Different segmentation scales (left: 100pixels, right: 50pixls).

We introduced NDVI (normalized differential vegetation index), NDWI (normalized difference water index), ENDVI (Enhanced Vegetation Index), Standard deviation near-infrared and other indexes to obtain the vegetation types. The computational formulas were as follows:

$$NDVI = {NIR - R}/{NIR + R} \tag{1}$$

$$NDWI = {G - NIR}/{G + NIR} \tag{2}$$

$$ENDVI = {R + G - 2 * B}/{R + G + 2 * B} \tag{3}$$

$$Red\_radio = {}^{(B + G + R + NIR)}\!\Big/_{R} \tag{4}$$

where, R is red band, G is green band, B is blue band, NIR is near-infrared band.

We constructed rule sets of vegetation classification by using formulas above (1) to (4) (Table 3), and constructed the Rule Sets (Table 3). Furthermore, the vegetation type such as dry farmland, bamboo forest was distinguished by vegetation texture point of Digital Line Graphic data.

Table 3. Rule sets of vegetation classification.

Land cover	Rule sets
Forest	NDVI > 0.4 and ENDVI < 0.1
Woodland	NDWI >−0.2 and green/blue < 1.45
Farmland	Standard deviation near-infrared <= 35
Other vegetation	NDVI > 0.4
Road	Red_radio<=6.5 and lend/with >=4 and ENDVI < 0.06 and NDWI <−0.02
Water	near-infrared <380 and NDWI> 0.05

It was clear that, object-oriented classification results seemed more homogeneous than that using spectral method which powerfully decreased salt-and-pepper noise. The boundaries between different vegetation types were very clear and results were accurate (Fig. 4). There are four types of vegetation (farmland, woodland, forest and other vegetation) based on eCognition Developer. If there was no assistant data, it was difficult for further subdivision of vegetation classes which was inconvenient for vegetation understanding and management. To obtain more types of vegetation, vegetation points were applied to subdivision.

2.3 Vegetation Polygons Optimizing Method

There were four steps of optimizing programs on vegetation polygons: assignment of vegetation attributes, normalization of segmentation boundaries, merge of porphyroclastic polygons and automatic topology modification. When attributes of vegetation points were assigned to segmentation polygons, the following principles should be followed: one polygon correspond with only one vegetation point, the vegetation point should be assigned to the polygon, one polygon correspond with two or more types of vegetation points, the polygon should be assigned mixed vegetation, the vegetation polygon with no points should keep the classified property. As the segmentation boundaries were irregularly complex surface and not consistent with DLG polygon, adaptive processing method was applied to smooth the edges according to VBA (Visual Basic for Applications) of ArcGIS Engine (Fig. 5).

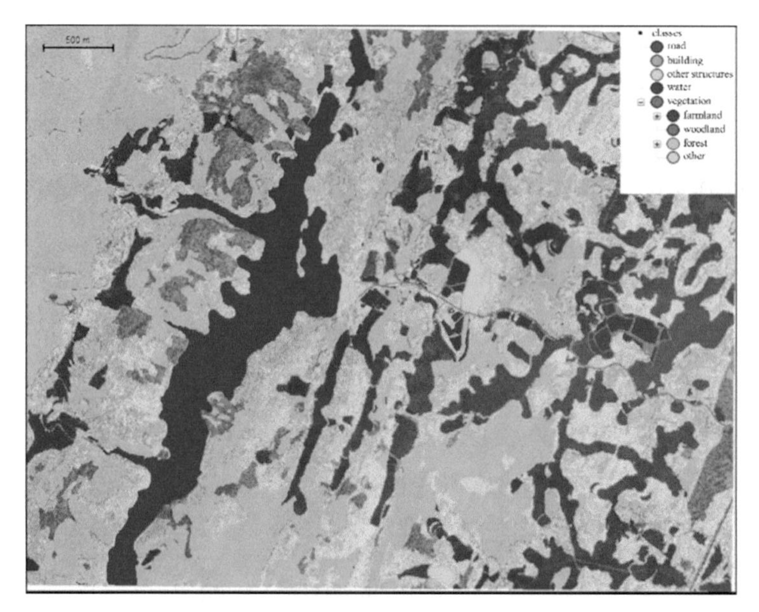

Fig. 4. Results of land cover classification based on eCognition Developer.

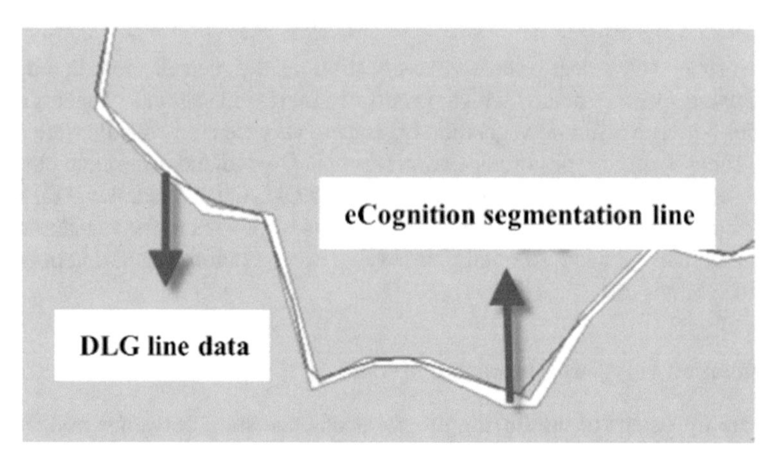

Fig. 5. Differences between DLG lines and smoothed vegetation boundaries

The boundary lines of DLG data were set as input parameters of buffer and buffer width was 0.8 m according to data features. Finally, the common side length between each buffer polygon and its near vegetation polygon was searched and calculated. The buffer polygons were merged to vegetation polygons of largest common edge lengths. The fusion program repaired the topological errors of and merged vegetation polygons with the same properties (Fig. 6).

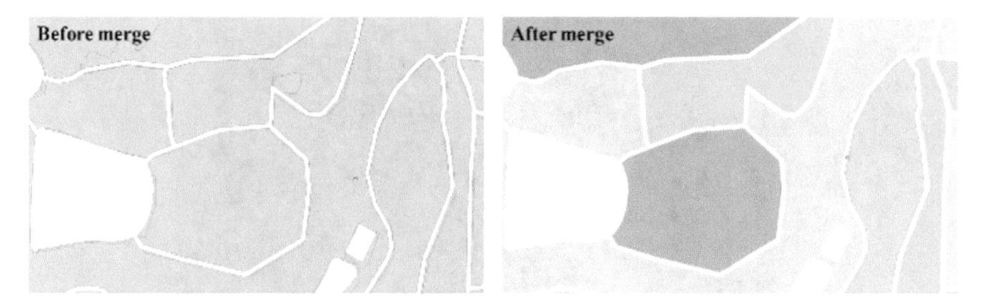

Fig. 6. Comparison between before and after merge of porphyroclastic polygons.

3 Evaluation of Construction of Vegetation Polygons

The data for evaluation was taking from manual interpretation of remote sensing image. Uncertain and interrogative vegetation classifications were carried out by field investigation. The result showed, the total number of automatically constructed vegetation polygons was 795, where correct types of vegetation polygons were 655 and false 140. The false types of vegetation polygons were rectified by manually adding guidelines. After 41 times of merge and 219 times of split according to guidelines, we obtained the final results of vegetation polygons (Table 4).

Table 4. Validation of construction of vegetation polygons.

Types	Before revised		After revised
Total number of vegetation polygons	795		926
Number of correct polygons	655		926
Number of false polygons	140		
	Merge times	41	
	Split times	219	
Accuracy (%)		82.39	71.73

It was concluded that, the number of real vegetation polygons was revised to 926. The accuracy of number of vegetation polygons before and after revised was 82.39%, 70.7% respectively (Fig. 7). The automatic construction and classification of vegetation polygon cost less than 10 h which was 8 times more than manual visual interpretation. It was obviously that, automatic method greatly improved the efficiency of production on vegetation polygons. The vegetation types were identified by vegetation texture points from Digital Line Graphic data of Fundamental Geographic Information.

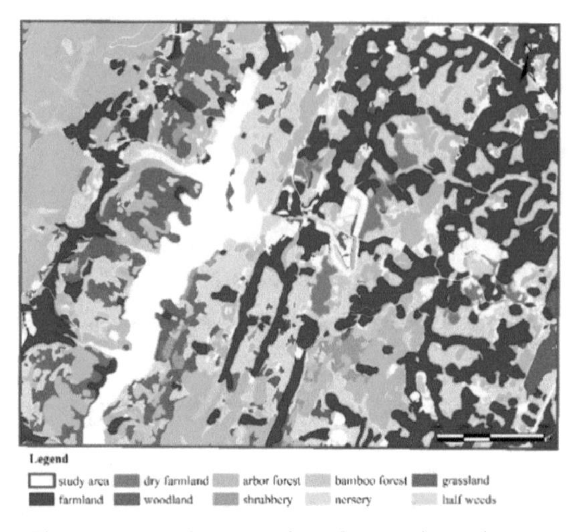

Fig. 7. Result of construction of vegetation polygons.

4 Conclusion

This paper introduced object-oriented method to vegetation classification which made full used of spectral texture, shapes and spatial location information of high resolution remote sensing image. The DLG data was utilized to multiresolution segmentation, and rule sets of classification were constructed for vegetation land cover based on eCognition Developer. The vegetation points were applied to subdivision of vegetation. Then, vegetation boundaries were normalized and smoothed, other processing algorithm was also used such as merge of small spots, topological modification and so on. The results were validated by field investigation and manual visual interpretation.

It was concluded that, object-oriented classification had more internal consistency and higher accuracy than traditional method that only using pixel information. Fast construction algorithm of vegetation polygons could increase efficiency significantly, with high accuracy and reliability. It provided a reference for batch production of vegetation classification.

However, we should attach weight to some problems in this process algorithm listed as follows.

1. The results of object-oriented classification largely depend on segmentation scales. In order to obtain the most suitable scale, it was necessary to take many times of tests which would cost some time.
2. The rule sets of land cover interpretation were suitable for the same sensor images which should be reset for other sensor images.

Acknowledgments. This work is jointly supported by Research and application of intelligent monitoring and evaluation of geological environment and restoration management of open-pit mines in Chongqing (Grant No. cstc2019jscx-gksbX0042).

References

1. Sansheng, C., Lifang, G., Liuhua, X.: Research on application of vegetation information based on remote sensing image in surveying and mapping. In: The 17th Conference on Remote Sensing of China (2010)
2. Vander, S.C.J., Jong, S.M., Roo, A.P.J.: A Segmentation and classification approach of IKONOS-2 imagery for land cover mapping to assist flood risk and flood damage assessment. Int. J. Appl. Earth Obs. Geo-inf. **4** 217–229 (2003)
3. Bao, C., Qiming, Q., Haijian, M.: Application of object-oriented approach to SPOT5 image classification: a case study in Haidian District, Beijing City. Geogr. Geo-Inf. Sci. **22**(2), 46–54 (2006)
4. Yunhao, C., Tong, F., Peijun, S.: Classification of remote sensing image based on object oriented and class rules. Geomat. Inf. Sci. Wuhan Univ. **31**(4), 316–320 (2006)
5. Xiaoxia, S., Jixian, Z., Zhengjun, L.: Extracting the river and the road using an object oriented technique from IKONOS panchromatic imagery. Sci. Surv. Mapp. **31**(l) 62–63 (2006)
6. Yan, H., Fengmin, W., Chao, Y., et al.: Research on typical features interpretation rule base in complicated mountain environment. Geospat. Inf. **14**(11), 9–11 (2016)
7. Xiang, F.: The discussion of some problems in large-scale DLG data production. Geomat. Spat. Inf. Technol. **36**(04), 186–188 (2013)
8. Xiang, F.: Research on establishing basic geographic information database based on MAPGIS and CASS. Henan Sci. Technol. **(03)**, 11–12 (2013)
9. Xuegang, M., Yao, Y., Shuxin, C., et al.: Forest stand identification based on e cognition software using QuickBird remote sensing image: a case of Jiangle forest farm in Fujian Province. J. Nanjing For. Univ. (Nat. Sci. Ed.), **43**(01), 127–134 (2019)
10. Rong, H., Hongjun, Q., Xuehu, W.: Application of eCognition in the interpretation of surface coverage in the census of geographical conditions. Bull. Surv. Mapp. (04), 134–135 (2014)
11. Yan, H., Ping, Q., Miao, J.: Ground objects extraction from high-resolution image based on eCognition in production. Geomat. Spat. Inf. Technol. **36**(12), 146–147 (2013)

Software and Information Systems

A/B Testing via Continuous Integration and Continuous Delivery

Ádám Révész[1] and Norbert Pataki[2(✉)]

[1] EPAM, Budapest, Hungary
[2] Department of Programming Languages and Compilers, Faculty of Informatics, Eötvös Loránd University, Budapest, Hungary
patakino@elte.hu

Abstract. Software version ranking plays an important role in improved user experience and software quality. A/B testing is a technique to distinguish between the popularity and usability of two quite similar versions (A and B) of a product, marketing strategy, search ad, etc. Nowadays, A/B testing can be utilized for the selection of preferred version of two akin software versions. Continuous Integration (CI) and Continuous Delivery (CD) are essential approach in modern software engineering. These approaches aim at the automated verification of the developed software. Moreover, automatic deployment of the software also can be supported. In this paper, we analyse how an automatic A/B testing can be realized in modern software development environments. We present an approach that supports long-term execution of A/B testing. We take advantage of widely-used CI/CD solutions and present an actual implementation.

Keywords: A/B testing · Continuous integration · Continuous delivery · Jenkins

1 Introduction

In the development lifecycle of web applications or web pages, most of the quality gates, such as static analysis checkers, unit tests and integration tests – to mention a few – are automated and part of the continuous integration (CI) and continuous delivery (CD) pipelines [1]. Most of the manual quality gates are also parts of these aforementioned processes [6]. These are working on multiple metrics we consider when we talk about quality in the likes of code coverage, code complexity, and of course testing application-specific business-related workflows [11]. Moreover, these pipelines enable the fast (re)deployment of the released software artifacts even in production environments [5].

In this user-facing application realm, a different aspect emerges, an aspect by which an application itself can boost business towards success or sentence it to failure. This crucial aspect is user experience (UX). UX can be developed intentionally with workflow enhancements, semantic user interfaces and following

S. Bourennane and P. Kubicek (Eds.): ICGDA 2022, LNDECT 143, pp. 165–174, 2022.
https://doi.org/10.1007/978-3-031-08017-3_15

coherent design languages, but measuring effectiveness of such developments can be done only by letting (mass) user-base get their hands on the new version of the application [13]. Of course we have a tool for measuring different metrics based on analytic data of application usage coming in from multiple live versions of the same application, A/B testing [8].

Timespans of A/B testing sessions are measured in days [14]. While this definitely manifests as a challenge, yet we decided to make efforts and state requirements against A/B testing systems and CI/CD systems with the purpose of integrating A/B testing as an automated quality gate into CI/CD pipelines, so most of the software development lifecycle is presented and trackable through CI/CD systems [9]. In this paper, we state these requirements, analyse general CI/CD system capabilities, design a high-level integration and present a proof of concept implementation. We take advantage of the widely-used Jenkins CI and its Blue Ocean for Jenkins pipelines [7].

The rest of this paper is organized as follows. In Sect. 2, we show how A/B testing works during software development. In Sect. 3, we propose an approach for the integration of A/B testing with CI/CD approaches. We present our implementation and its details in Sect. 4. Finally, this paper concludes in Sect. 5.

2 A/B Testing

A/B testing is a dynamic test method to use when multiple versions of the same application have multiple viable versions – in general form version A and version B, hence the naming – and the decision between them is to be made based on metrics which need runtime collected analytic data from the target audience [12].

A working example is a webshop where version A and version B have different user interface (UI) for handling basket. The metrics under examination could be any variation of the following few examples:

- count of added items from suggestion list on basket UI
- time spent on basket ui before proceeding to payment
- count of clicking on basket items (for item details) on basket UI
- count of unfinished orders – customers have not proceeded from basket UI.

2.1 General Characteristics of an A/B Testing System

Configuration. An A/B testing system is configurable, it preferably supports multi-project configuration. For this purpose, an optional web interface is available in forms of web API and web UI. Project configuration contains - amongst implementation-dependent meta properties – project id, metrics to evaluate with implicit (built-in) and explicit queries.

Collecting Analytic Data. An A/B testing system contains analytics listener service which receives runtime analytic data from client instances (from users) and stores them in a database which is preferably indexed to support general queries used on evaluation (by project id, event datetime, implementation-dependent event meta).

An A/B testing system often exposes web API, web-socket or other general interfaces on which clients can send analytic data to analytics listener service. Optionally, A/B testing systems publish Software Development Kits (SDKs), libraries for their most common target platforms (e.g. Android, iOS, JavaScript, Xamarin) which eases client configuration and wraps analytics collecting – sending implementing platform-specific best practices.

Test Session Scheduling. An A/B testing system has a service for scheduling A/B test sessions [8]. Test sessions can be scheduled with project key, A- and B versions to compare and timespan of the session. Scheduled tests are listable, readable, cancellable, winner can be announced manually.

Test Evaluation. An A/B test system has query service which runs on-demand and scheduled queries evaluating test sessions. Query service shares database with analytics listener service.

3 Integration

As stated, main goal is to integrate A/B testing systems into CI/CD pipelines as an automatic quality gate in the flow of software development lifecycle.

3.1 System Requirements

General CI/CD system requirements for integrating A/B test systems are the following:

- Ability to trigger job (running instance of an actual pipeline) on web API call (webhook)
- Pipeline call external web services through HTTP.

General A/B test system requirements for integrating with CI/CD systems:

- Ability to store CI/CD system and pipeline metadata per-project basis
- Ability to call webhooks on A/B test events

Project pipeline metadata contains template for making webhook call, triggering pipeline, and pipeline-specific variable mappings.

3.2 Event-Driven Design

Ability to call webhooks fairly can be stated general capability of any current CI/CD tool but determining meaningful events to listen to is when integrating is far more interesting [3]. A/B test sessions have set timeframe to examine. Having this in mind, end of session could be the main event. If we consider that these A/B test sessions are triggered from CI/CD pipelines and even the timespan arrives from CI/CD system as a parameter, session start event gets ruled out as meaningful event to be propagated back to CI/CD systems. Before eagerly stating end of session is the only event, we have to consider the capability of manual session termination. Depending on chosen abstraction, these can be treated as separate events or can be differentiated with event parameter. Other A/B test system internal events such as analytic data arrival, project configured, session scheduled are not interesting integration-wise.

A/B Test Event Details. Considering the followup stages of automated pipeline, the manual or scheduled nature of the session end event can be omitted. Integration point of view the following parameters are highly important for passing the test results to pipeline call:

- `decidable: Boolean` – Indicating if the query evaluation result clearly tells whether A- or B version is the winner
- `winner: String` – Exact version of the winner

For enriching pipelines – e.g. for log or visualisation purposes - extra arguments can be passed like:

- `isManual: Boolean` – Indicating whether termination is user-triggered event
- `user: User` – Metadata of user who triggered session end if manual
- `resultset: Dictionary` – Query result for compared versions per-metric

Event Handling. Since A/B testing happens in production environments optional manual approval can be put before each step when handling end of A/B test session event. Focusing on key steps, such approvals are omitted in the following discussion.

Because services and applications can be published, distributed, configured many ways, discussion on handling their scenarios of terminating loser- and rolling out winner version would go on epic lengths. In Sect. 4, an actual example is discussed.

Deploying A- and B versions of an application can happen by distributing actually different application artifacts (service images, application bundles, etc.) or using feature toggles (also known as feature flags). Feature flags enable the same artefact to run version A and B as well, but they are controlled through remote configurations. As a result, no matter which way the A/B testing is prepared, in the end, a repository (containing either application of configuration code) has to reflect the decision over the winner version. The loser version of the application or configuration has to be terminated and replaced with winner.

4 Implementation

4.1 Proof of Concept Environment

Running environment is a Kubernetes cluster, since Kubernetes is an industry standard container orchestration system [2]. It enables this proof of concept to be easy to reproduce on various setups (from single machines to cloud providers) while gives abstraction tools to clearly separate resources for each actors in our example: A/B test system, CI/CD system, test subject.

Subject project of A/B testing is a simple webpage containing a button with different label for A and B version. Application source code is version controlled on a remote git server. The application name in Kubernetes is "testApp". Kubernetes service "testApp" balances load by round robin strategy between "testApp" instances.

A/B testing system is based on our own solution introduced in the paper [8]. We discuss our improvements later.

CI/CD system is a Jenkins instance, configured to have read and write access to git repository, has read and write access to Kubernetes deployments on "test" namespace.

4.2 A/B Test System

In this section, we describe the high-level architecture of our POC A/B Test System with its main components in the same order as they have been discussed in Sect. 2. Figure 1 presents the architecture of our approach.

Configuration. Configuration is stored in Kubernetes Config Map resource. Configuration contains a Lucene query string and a webhook template with following variables for each project:

- decidable
- winner

For POC purposes, exposing configurations through any interface is not required, config maps can be handled with ease with Kubernetes CLI tool.

Collecting Analytic Data. Listener service could be omitted from POC and data could be sent directly to ElasticSearch LogStash Kibana (ELK) stack, but it plays important semantic role. Listener service is a simple REST API, which accepts POST requests on "/data" endpoint with JSON payload requiring the following fields:

- `projectKey`: `String`
- `version`: `String`
- `eventType`: `String`, Enum: "`click`" – the single event supported in this POC
- `eventMeta`: `Object`

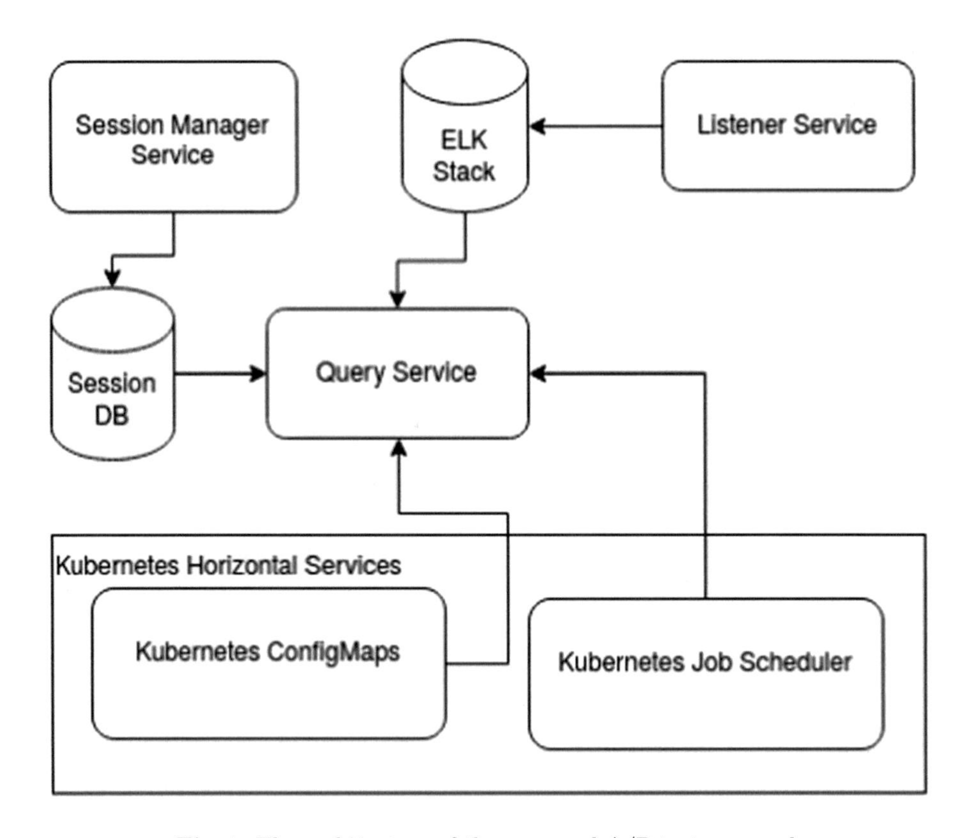

Fig. 1. The architecture of the proposed A/B test approach

Listener service formats the payload and forwards it to the ElasticSearch LogStash Kibana stack.

Having Listener service enables us not to directly expose ELK stack outside the Kubernetes cluster.

Test Session Managing. Session Manager exposes "Schedule" entity (stored in Postgres DB) through REST API with the following properties:

- `projectKey: String`
- `versionA: String`
- `versionB: String`
- `start: DateTime`
- `end: DateTime`
- `completed: Bool`
- `winner: String`

Note, schedule can be manually completed through REST API.

Test Evaluation. Proof of Concept does not support on-demand peeking of temporary results of ongoing A/B test sessions. Query service is implemented as a Kubernetes Cron job which starts up every 10 min. Query service

- loads project configurations on startup
- checks schedule table for incomplete past due session
- runs project query with actual A- and B versions from current schedule
- triggers project webhook substituting variables with actual values from schedule object and query result
- marks schedule object completed

4.3 Example Scenario

In this section, we present an example Jenkins pipeline with integrated A/B test. Pipelines introduced here are simplified in such terms these do not contain stages like unit testing, deployments to lower environments, since A/B test run on production on close to final versions. The following scenario assumes GitFlow is being followed where release branches are created from develop branch.

Pre-A/B Test Section. Following test initialisation pipeline is illustrated on Blue Ocean Jenkins UI that can be seen on Fig. 2. The pipeline consists of the following steps:

- Jenkins pipeline triggered on commits to release/vX.Y.Z branches
- Application gets deployed to Kubernetes cluster to "test" namespace keeping old replicaset active as well (so earlier version keeps running)
- Lookup latest git tag with "-stable" postfix
- Schedule A/B test session, providing project key, latest "-stable" version as version A X.Y.Z version as B, setting preconfigured timespan starting with current time
- Exit

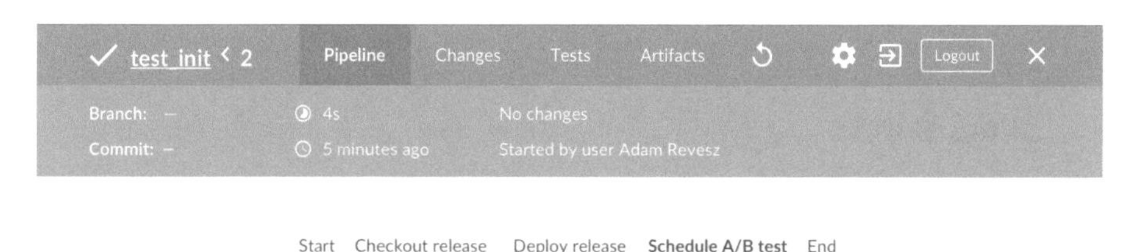

Fig. 2. Initialisation pipeline

Note that pipeline job has to be terminated since A/B tests usually take days and having pipeline runners actively wait that long is not advised. Current popular CI/CD systems lack the functionality of "waking up" pipeline jobs through webhooks, asynchronously [4].

A/B Testing. Application in use, logs generated.

Post-A/B Test Section. Following test conclude pipeline is illustrated on Blue Ocean Jenkins UI that can be seen on Fig. 3. It consists of the following steps:

- A/B test system triggers post-A/B Test pipeline
- In case of winner can be determined
  - In case of A (older) version is the winner: B version gets terminated
  - In case of B (new) version is the winner: A version gets terminated. B version gets tagged with "X.Y.Z-stable". With manual approval release/v.X.Y.Z branch gets merged to main branch.
  - With manual approval release/vX.Y.Z branch gets deleted
- In case of winner cannot be determined: SKIP

Jenkins, due to its core design does not support static pipeline processing which results in disappearing of conditional stages not run in latest job [10]. E.g. on Fig. 2, only the execution of only one case is shown, when winner can be determined and B version won.

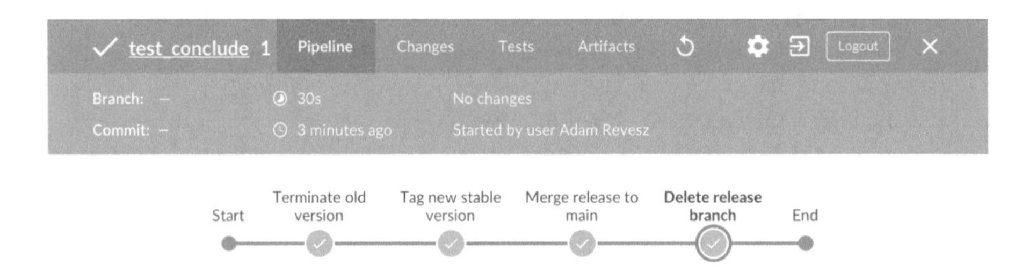

Fig. 3. Test conclude pipeline

5 Conclusion

CI/CD solutions are widely-used in modern software development. These solutions provide fast feedback to the developers whether the source code of software artifacts is proper and enable the deployment of the application. In this paper, we have stated our requirements agains both CI/CD and A/B testing systems in order to successful integration. We have presented proof of concept design and implementation of such A/B testing system and CI/CD pipelines.

Starting A/B tests sessions, running and evaluating tests and even followup tasks such as version tagging, code repository handling, loser version terminating can be done without manual intervention. While these are definitely satisfying results, note has been taken on why current CI/CD systems fail to contain long-running (outer) tasks in a single pipeline, making a step harder understanding the whole workflow as a unit. Experimenting with system specific meta-data passing might be beneficial so post-A/B test pipeline could be shown as down-stream job of pre-A/B test pipeline job, but this functionality is not yet supported in general.

References

1. Armenise, V.: Continuous delivery with Jenkins: Jenkins solutions to implement continuous delivery. In: 2015 IEEE/ACM 3rd International Workshop on Release Engineering, pp. 24–27 (2015). https://doi.org/10.1109/RELENG.2015.19
2. Bernstein, D.: Containers and cloud: from LXC to Docker to Kubernetes. IEEE Cloud Comput. **1**(3), 81–84 (2014)
3. Chen, L.: Continuous delivery: huge benefits, but challenges too. IEEE Softw. **32**(2), 50–54 (2015). https://doi.org/10.1109/MS.2015.27
4. Fitzgerald, B., Stol, K.J.: Continuous software engineering and beyond: trends and challenges. In: Proceedings of the 1st International Workshop on Rapid Continuous Software Engineering. RCoSE 2014, pp. 1–9. Association for Computing Machinery, New York (2014). https://doi.org/10.1145/2593812.2593813
5. Lehtonen, T., Suonsyrjä, S., Kilamo, T., Mikkonen, T.: Defining metrics for continuous delivery and deployment pipeline. In: Nummednmaa, J., Sievi-Korte, O., Mäkinen, E. (eds.) Proceedings of the 14th Symposium on Programming Languages and Software Tools (SPLST), pp. 16–30. No. 1525 in CEUR Workshop Proceedings, Aachen (2015). http://ceur-ws.org/Vol-1525/paper-02
6. Oberle, T., Szabó, C.: An architectural prototype for testware as a service. In: 2015 IEEE 13th International Symposium on Applied Machine Intelligence and Informatics (SAMI), pp. 15–19 (2015)
7. Pathania, N.: Declarative Pipeline Development Tools, pp. 191–209. Apress, Berkeley (2019). https://doi.org/10.1007/978-1-4842-4158-5_5
8. Révész, Á., Pataki, N.: Containerized A/B testing. In: Budimac, Z. (ed.) Proceedings of the Sixth Workshop on Software Quality Analysis, Monitoring, Improvement, and Applications, pp. 14:1–14:8. CEUR-WS.org (2017). http://ceur-ws.org/Vol-1938/paper-rev.pdf
9. Révész, A., Pataki, N.: Continuous A/B testing in containers. In: Proceedings of the 2019 2nd International Conference on Geoinformatics and Data Analysis, ICGDA 2019, pp. 11–14. Association for Computing Machinery, New York (2019). https://doi.org/10.1145/3318236.3318254
10. Révész, A., Pataki, N.: Visualisation of Jenkins pipelines. Acta Cybernetica (2021). https://doi.org/10.14232/actacyb.284211, https://cyber.bibl.u-szeged.hu/index.php/actcybern/article/view/4119
11. Schaefer, A., Reichenbach, M., Fey, D.: Continuous Integration and Automation for Devops, pp. 345–358. Springer, Dordrecht (2013). https://doi.org/10.1007/978-94-007-4786-9_28

12. Tamburrelli, G., Margara, A.: Towards automated A/B testing. In: Le Goues, C., Yoo, S. (eds.) SSBSE 2014. LNCS, vol. 8636, pp. 184–198. Springer, Cham (2014). https://doi.org/10.1007/978-3-319-09940-8_13
13. Török, M., Pataki, N.: Service monitoring agents for DevOps dashboard tool. In: Heričko, M. (ed.) Proceedings of the 21th International Multi-Conference INFORMATION SOCIETY IS 2018, Volume G: Collaboration, Software and Services in Information Society, pp. 47–50 (2018)
14. Xu, Y., Chen, N., Fernandez, A., Sinno, O., Bhasin, A.: From infrastructure to culture: A/B testing challenges in large scale social networks. In: Proceedings of the 21th ACM SIGKDD International Conference on Knowledge Discovery and Data Mining. KDD 2015, pp. 2227–2236. Association for Computing Machinery, New York (2015). https://doi.org/10.1145/2783258.2788602

An Automated Approach for Mapping Between Software Requirements and Design Items: An Industrial Case from Turkey

Selin Karagöz[1,2](✉) and Ayşe Tosun[2]

[1] ASELSAN A.Ş., Ankara, Turkey
skaragoz@aselsan.com.tr
[2] Faculty of Computer and Informatics Engineering, Istanbul Technical University, Istanbul, Turkey

Abstract. When a new request comes to the existing software, determining whether there will be reuse and determining where the new requests will be mapped in the existing design are important problems. Since this process is done manually by developers in the context we work, it depends on experience and domain knowledge, besides it is an error-prone and time-consuming process due to the human factor. The main purpose of this study is to correctly predict which new requests in the System Design Document (SDD) match which feature set in the existing software's Software Requirement Specification (SRS) document. We consider the feature mapping problem between SDD items and SRS requirements as a multi-label multi-class classification problem. Zemberek, a Turkish natural language processing library, is used for preprocessing and feature extraction of the SRS document of the existing software and three SDD documents of different systems to which this software will be delivered. The features extracted from the SRS document are categorized under a certain number of feature topics using the LDA algorithm. The FastText algorithm and AdaBoost-based classifier ICSI-Boost are used to decide which of the topics from the SRS document represents a feature in the SDD document, and the predictions are compared with manually determined topics by experts. ICSIBoost achieves quite 67% to 90% precision in topic predictions, whereas the FastText algorithm does not meet our expectations for small and imbalanced data.

Keywords: Software product line · Topic modeling · Feature mapping · Turkish NLP · Multi-label multi-class classification

1 Introduction

A software product line is defined as "a set of software-intensive systems sharing a common, managed set of features that satisfy the specific needs of a particular market segment or mission and that are developed from a common set of core assets in a prescribed way" [1]. Software product lines (SPL) ensure the quality, and manageability of software system families [2]. A product in software product line consists of reused core

S. Bourennane and P. Kubicek (Eds.): ICGDA 2022, LNDECT 143, pp. 175–186, 2022.
https://doi.org/10.1007/978-3-031-08017-3_16

assets and product-specific custom assets [3]. Building SPL with reusable components is critical to maintain code quality and to manage project schedules shorter and easier [4]. Reusable components are very difficult to identify if the system is not designed with proper documentation and traceability constraints. There are a few methods in the literature to automatically identify these components, such as search-based techniques [5, 7] or object oriented model-based extractions [6, 7].

In software companies, if the reusable components are not determined in an automated way or are not documented beforehand, there is a serious need for manual effort to identify those in later revisions of the software product. With each new project request, all documents need to be reviewed by the developers, and feature matching can be made between these documents. Especially in industrial contexts that develop safety-critical systems, the inability to match new requests to existing features causes delays in final project delivery. In this study, we also aim to identify reusable requirements of a safety-critical software product line to address customer needs over multiple releases. The capabilities of the software system developed in our industrial context are determined by developers through a manual examination of the System Design Document (SDD) and Software Requirement Specifications (SRS) Document. The requirements in the SRS document represent the core assets, whereas the items in the SDD documents represent the custom assets. SDD defines the design of the system that will meet the user requirements and includes design decisions made at the system level. In addition, basic design information about all hardware and software components as well as general system architecture are described in this document. SRS document has a narrower scope than SDD document, and contains all the user requirements of a software component. When a new project is added to the product line, the items of a software component in the SDD of the project are examined by the developer who is responsible for the related software. Then, these items from the SDD document are manually matched with the relevant requirements in the SRS document of the current software. This software can be given directly to the project if all the features have already been defined in SRS, and hence, implemented in the current software. If there are requirements that cannot be matched, these requirements are considered as new features and included into the development plan in accordance with the project schedule. If a feature that is in the SDD of one project, but not in the SDD of another project and if that feature is not among the core assets of the existing software, a project-specific feature is considered. The use cases of these features in the user interface are set via configuration files. Currently, the matching process is mostly done with manual effort.

To tackle this problem, we propose an automated approach for mapping SDD items to SRS requirements on the software that is part of a SPL developed in our industrial setting for the defense industry. Although most of the capabilities offered by the software are commonly used by the customers, some features may vary according to the requirements on different platforms. We aim to predict which feature in the SRS document matches with a new requirement in the SDD document. Since each SDD item can be mapped to one or more SRS requirements, we define our problem as multi-label multi-class classification problem. SDD and SRS documents are prepared in Turkish as they contain the features of the software to be used in Turkey. We use Zemberek, a Turkish natural language processing library, to preprocess the documents in the study. The requirements in the SRS document are clustered into 5 topics using LDA. Each requirement is assigned to one or more topics. The requirement topics to which each item in the SDD document is assigned are determined manually by the experts in the company. Then two classification algorithms namely FastText and ICSIBoost, are utilized to predict which requirement topics that the SDD items are assigned to. The performance of the algorithms are assessed against expert-based topics of SDD items.

This paper is organized as follows. Section 2 provides the literature review on identifying reusable components of a software product and feature extraction from documents. Section 3 expresses the steps of our methodology in detail. Section 4 discusses the results of the used algorithms in the study. Finally, Sect. 5 presents the conclusion and future work.

2 Related Works

In SPL, a common problem is to detect if a new request matches any feature in an existing feature set. It is important to determine the dependencies between these features and to develop by reuse after determining them. Identifying reusable components of a software product has been a challenge that was addressed by several studies in the literature [3, 5, 6, 8, 9]. The study in [8] presents the novel models and the reusability-metric from the component-based system for all categories of components. Also, the selection criteria are presented for components by using the reusability features of component-based software in this study. Another study models the Reusable Software Component Identification problem as a search-based multi-objective problem [5]. This problem models a software system using object-oriented source code to identify optimized reusable software components. Another study considers the interactions between client applications and the targeted API and proposes an approach to identify reusable software components in object-oriented APIs [6]. The study in [7] aims to identify the components from an existing software system and transform those components to the Java Bean component model by utilizing a search-based PSO algorithm and clustering the software system. One study analyzes the artefacts that need to be configured during SPL [3]. The authors present an evolution-based configuration management model used for products in a SPL that can easily extract core parts and custom parts. Another study proposes a tool namely MoSPL that handles version management at the component level via its product versioning and data models [9].

In this study, we utilize design documents for feature extraction. There are different approaches in the literature for extracting features from Software Requirements documents. Some studies have used Part of Speech (POS) tagging [10]. Hamza and Walker [11] present the FFRE method that assists the extraction of features and the relationships between these features from requirements documents. FFRE creates feature models from SRS documents for SPL-based systems using natural language processing (NLP) techniques [11]. Another study presents a fully automatic approach to extract software requirements from design documents as well as to compare the extracted requirements to those that exist in the official software requirement database [12]. Although the study in [12] becomes an inspiration for us, we conduct a different analysis to cluster requirements in SRS and match those which the items in SDD. In our context, there is no method to match a new incoming request to an existing feature using the document classification method, so we decide to propose an automated technique to be able to match those more efficiently and accurately.

3 Methodology

3.1 The Project Under Study

In our industrial setting, every software component has its own SRS document containing a full set of capabilities. In addition, every project, which represents a system that includes many software and hardware components, has an SDD document that contains the basic requirements for all components in the project design. The requirements for each software component are found under the heading of the relevant software in the SDD document. The developer examines the items under this heading by considering the requirements in the SRS document. If an item is an existing requirement included in the SRS document and is met by the existing software, there is no need for additional development. However, if it cannot be matched with any item in the SRS document, it is added to the development plan as a new capability. Also, an item from the SDD document can match more than one item in the SRS document. For this reason, we deal with our problem as a multi-label.

In this study, we use one SRS document belonging to a safety-critical software and three related SDD documents of three projects developed based on this software. Since the SDD document includes the entire system design, the scope of the requirements for each component is narrower and includes the most essential capabilities needed. Therefore, the number of new requirements for the relevant software taken from the SDD document is considerably less than the number of requirements defined in the SRS. The number of requirements of the relevant software in the mentioned documents is given in Table 1.

Table 1. The number of software requirements in the documents

Document name	Number of software requirements
SRS	560
SDD-1	22
SDD-2	17
SDD-3	9

3.2 Preprocessing

For language processing, the methods and classes of the Zemberek library [13] developed by Tubitak for Turkish natural language processing are used in this study. The "TurkishMorphology" class is used for word analysis and production. The "TurkishTokenizer" class is a rule-based tokenizer. The "TurkishSpellChecker" class checks if the spelled words are spelled correctly, and if there are any mistakes in the spelling, it offers suggestions to correct them.

First of all, the Turkish stop words list in the documents has been determined. While roots are found, stop words in this list are ignored. The list of Turkish stop words determined in the context of this study is shown in Fig. 1.

A	B	Ç	H	N	Ö	Ü
acaba	bazı	çok	hem	ne	önce	üzeri
ait	belki	çünkü	hep	neden	**S**	üzerine
ama	benzer	**D**	hepsi	nerede	sonra	üzerinde
anda	bin	da	hiç	nereye	**Ş**	üzerinden
arasında	bir	daha	**İ**	niçin	şey	**V**
arasındaki	biri	dahi	için	niye	şeyden	vb.
aslında	birkaç	de	ile	**O**	şeyi	ve
aşağı	bir şey	defa	ilgili	olan	şeyler	veya
aşağıda	bir şeyi	diye	ise	olanlar	şu	vs.
aşağıdaki	bu	**E**	**K**	olarak	şuna	**Y**
aynı	buna	eğer	kendi	ona	şunda	ya
az	bunda	en	kendisine	ondan	şundan	yani
	bundan	**G**	kez	onlar	şunu	ya da
	bunu	gibi	ki	onlardan	**T**	yeni
	bunun	göre	**M**	onları	tüm	yukarı
			mı	onların		
			mu	onu		
			mü			

Fig. 1. The list of Turkish stop words determined in the context of this study

Since the documents are written using a formal, proper language, there is no need for normalization during pre-processing. Hence, only tokenization and word stemming are applied to the entire text of the SRS document and the requirement texts retrieved from the related software in SDD documents. During the tokenization process, punctuation marks and numbers are also cleared from the text. Then, each sentence is expressed as a list of tokens. Since these tokens are analyzed as roots, the stemming of words is considered as the most basic preprocessing step in this study.

While finding the roots from these tokens, the Zemberek object is created with Turkish selection as a language option. The kelimeDenetle() method checks whether a word is spelled correctly. If the word is Turkish, this method returns true. Then, the word class is made into a list with the kelimeCozumle() method. The root of the word is obtained with the kok() method. Since the words obtained from the document are recorded as String type, the icerik() method is used to transform the value in the root object.

We ensure that at least 3-letter roots are taken, while we eliminated 2-letter roots since they disrupt the semantic integrity. TF-IDF values are calculated for all the words in the document, and the words with the same roots are compared according to these values, so that root selection is made.

3.3 Topic Modeling

As the next step, we apply topic modeling only to the pre-processed SRS documents for extracting the feature sets using the LDA algorithm. Determining the correct number of topics for topic modeling is an important problem. There are different studies in the literature for finding the correct number of topics [14–16]. In this study, perplexity is used to determine the number of topics. Perplexity is a measure used to evaluate the success of a model in describing a dataset. The perplexity value should be lower for a better model [17]. We applied two different data splitting strategies (5-fold cross validation and 90%-10% split) to see if the word distribution is consistent on 560 requirements in the SRS document according to the perplexity values.

First, 5-fold cross-validation is applied and it is seen that the perplexity values calculated for each fold are quite close to each other. Then, 500 items are selected as the train, whereas 60 items as the test group, and random sampling is applied and this process is repeated 10 times. Perplexity values calculated as a result of this cross-validation are also found to be very close to each other. Then, perplexity values are calculated for the 5-10-15-20-25-30-35-40-45-50 topics. The calculated values are averaged and compared with each other. It is observed that the lowest perplexity values are obtained for five and 10 topics. As a result, it is evaluated that the number of topics could be between 5 and 10, and a recalculation is made for this number of topics. However, since the lowest value is taken for 5 topics, it is decided to use 5 topics. The perplexity values calculated for different number of topics are listed in Table 2.

Table 2. Perplexity values for different number of topics

Number of topics	Perplexity values
5	6,32
6	7,21
7	8,16
8	8,46
9	9,28
10	10,59
15	15,18
20	16,65
25	20,52
30	21,25
35	18,47
40	19,72
45	18,11
50	19,64

In order to purify the verbs that are frequently used in the requirement clauses from the phrases contained in the topics, these verbs (sun, sağla, çalış, gel, gerçekle, bulun, kullan, başlat) are also ignored and the final topics are found. The final topics are shown in Fig. 2.

Fig. 2. Word distributions included in the topics

The topics' suitability for each of the SRS requirements is evaluated according to the topic probabilities calculated for each item, and the items are assigned to more than one topic for similar probabilities. Analyst approves that the topics cluster requirements in a semantically meaningful way.

3.4 Matching Requirements with SDD Items

After the topics are found associated to the SRS requirements, we matched the SDD items with these topics. The items in the three SDD documents are reviewed and an expert-based evaluation was made to determine which items correspond to which requirements in the SRS document. Then, the topics that match the requirement in the SRS document are assigned to the matched SDD items. If an SDD item matches more than one requirement in the SRS, and each of those requirements matches more than one topic, more than one topic can be assigned to that item. In such a case, the probability of being assigned to irrelevant topics may increase as well. Therefore, we calculate an impact ratio for topics assigned to each SDD item. The impact ratio of a topic for an SDD item is calculated in Eq. (1). In this formula, requirements represent SRS requirements that match with the SDD item.

$$Impact\ ratio\ of\ a\ topic\ for\ an\ SDD\ item = \frac{Number\ of\ requirements\ assigned\ to\ the\ topic}{Number\ of\ requirements} \quad (1)$$

For example, an SDD item can match five requirements in the SRS document. If four of these requirements match with Topic-1 and one of them matches Topic-2, then the item is assigned to both Topic-1 and Topic-2 although the impact of one is greater than the other. We set a threshold value of 20% to decide which of these topics would be the final categories of an SDD item. Based on the impact ratios of both topics (80% for Topic-1 and 20% for Topic-2) and the threshold value, we assign only Topic-1 to this item. The distribution of SDD items to topics are given in Fig. 3. Since Topic-5 cannot be adequately represented in this very limited data, an estimation on this is impossible. For this reason, we do not consider Topic-5 and set the final number of topics as 4. Since we aim to classify items with these topics, we consider the problem as multi-class and multi-label.

The number of SDD items	48
Average number of requirements matching an SDD item	5.5
Average number of topics matching an SDD item	1.6
The number of SDD items assigned to Topic-1	22
The number of SDD items assigned to Topic-2	18
The number of SDD items assigned to Topic-3	16
The number of SDD items assigned to Topic-4	16
The number of SDD items assigned to Topic-5	3

Fig. 3. Some statistics of the data after the matching process

3.5 Classification Algorithms

We first employed FastText [18], a text classification algorithm that supports multi-label multi-class classification to train the model. FastText is a word embedding-based text classifier developed by Facebook. It is chosen because it is a fast and recent approach. Since FastText works with very large datasets in tag prediction experiments [18], we decided to employ another, more typical classifier from machine learning. Then, ICSI-Boost, an open-source implementation of the Adaboost-based classifier BoosTexter [19], is used as another method to train the model. When the datasets used in two studies [18, 19] are compared, it has been observed that BoosTexter works with smaller data than FastText.

FastText and ICSIBoost are used to predict the topics of the items in the SDD document using the topics assigned based on the SRS documents. The SDD items are divided into approximately 90% train and 10% test set. Since there are only 48 SDD items in total and these items are distributed unevenly over 4 topics, an iterative stratification method [20] is applied. Accordingly, items are distributed to the train and test sets starting from the least number of topics. The data is divided into 3 folds, i.e., 16 items in each fold are used for testing, whereas the remaining 32 items are used for training and validation. The model is trained with the train data determined for each fold, and assessed on the test data.

3.6 Performance Evaluation

The results of our multi-label, multi-class classification are reported in 4 × 4 confusion matrices. The confusion matrices are created by comparing the topic labels assigned to the SDD items with the actual topics assigned by the expert. In the matrices, the rows represent the predicted classes, whereas the columns represent the actual classes. If an assigned topic for each item could be predicted by FastText or ICSIBoost, that topic value is incremented by 1. If another topic is predicted instead of the actual one, the predicted topic value is increased by 1. Based on this confusion matrix, recall, precision, and F1-measure values are computed to assess the model's prediction performance [21].

The precision rate for each topic is calculated using the Eq. 2. While calculating the precision, we consider the number of items assigned to a topic as the TP value. For a topic, we take the number of all SDD items estimated by FastText or ICSIBoost belonging to that topic as the TP+FP value. Then we calculate the precision values for each topic. We made such a calculation in order to obtain the optimum result for the precision values. Recall refers to the actual classes predicted correctly and it is calculated using the Eq. 3. F1-Score value refers to the harmonic average of precision and recall values. F1-Score is calculated using the Eq. 4.

$$Precision = \frac{TruePositive}{(TruePositive + FalsePositive)} \tag{2}$$

$$Recall = \frac{TruePositive}{(TruePositive + FalseNegative)} \tag{3}$$

$$F1 - Score = 2 \times \frac{Precision \times Recall}{Precision + Recall} \tag{4}$$

4 Results and Discussion

Here, we present and discuss the prediction performance of FastText and ICSIBoost over all the topics. Tables 3 and 4 show the confusion matrices of FastText and ICSIBoost predictions, where the rows represent the predicted classes and the columns represent the actual classes, respectively. Table 5 lists the precision, recall, and F1-score rates of FastText.

Table 3. Confusion matrix of FastText

Topics	Topic-1	Topic-2	Topic-3	Topic-4
Topic-1	18	6	5	5
Topic-2	3	7	5	3
Topic-3	5	11	10	7
Topic-4	6	16	13	13

Table 4. Confusion matrix of ICSIBoost

Topics	Topic-1	Topic-2	Topic-3	Topic-4
Topic-1	19	5	4	4
Topic-2	5	10	6	6
Topic-3	2	6	8	4
Topic-4	3	6	6	12

Table 5. Performance of FastText

Topics	Precision	Recall	F1-Score
Topic-1	0.75	0.56	0.64
Topic-2	1	0.18	0.30
Topic-3	0.59	0.30	0.40
Topic-4	0.45	0.46	0.46

Although the precision values of Topic-1 and Topic-2 are good, we could not achieve the desired success in recall values, since this is a difficult problem. If a random classifier were used in a 4-class problem, 25% accuracy could be predicted. According to our recall values, we observe a better success than the random classifier in the 1st, 3rd, and 4th topics. However, this is not a practically applicable model for us. Table 6 lists the precision, recall, and F1-score rates for predictions of ICSIBoost.

Table 6. Performance of ICSIBoost

Topics	Precision	Recall	F1-Score
Topic-1	0.90	0.66	0.76
Topic-2	0.71	0.37	0.49
Topic-3	0.67	0.33	0.44
Topic-4	0.75	0.46	0.57

According to the precision values of ICSIBoost, we observe a better success than the FastText in the 1st, 3rd, and 4th topics. Also, we get better results for all topics according to recall and F1-Score values with ICSIBoost compared to FastText. The fact that the data we have is very limited and the distribution of the items to the topics is unbalanced, is one of the major reasons that reduce the success of both algorithms. It could not be tested on another data set since the problem is domain-specific.

Another reason may be the number of topics. Since the SRS document is not divided into topics, it was necessary to cluster the requirements according to a certain number of topics. First of all, we determined the number of topics is 25, considering the number of requirement headings in the SRS document. However, we could not get consistent results with 25 topics. Then, we calculated the perplexity values for various topic numbers and determined the number of topics as 5 according to the results. We analyzed the requirements that we clustered into 5 topics and evaluated that the topics were semantically meaningful. However, Topic-5 could not be adequately represented in this very limited data we do not consider Topic 5 and set the number of topics as 4. We evaluated that we could get better results if we chose fewer topics. However, we did not prefer to choose a fewer number of topics because the topics were separated in a meaningful way.

5 Conclusion

When a new request come for the existing software, developers review the documents containing the software requirements and check if an existing feature matches with the new request. Thus, they decide if there is reuse or not. Since this process is done manually by developers, it depends on experience and domain knowledge. Also, it is an error-prone and time-consuming process. Automating the processes of matching capabilities between documents, identifying common or different capabilities using algorithms on the backend shortens the project development process and reduces the margin of error. It also contributes greatly to the development and growth of the software product line.

In this study, we consider the problem of automatically matching the SDD items and the requirement groups obtained from the SRS document as a multi-label and multi-class classification problem. First of all, topics found manually are compared with topics found with FastText. 75% precision value is achieved for Topic-1, and 100% precision value is achieved for Topic-2. We also get better results for Topic-3 and Topic-4 than random classifier for 4 classes. However, we could not achieve the desired success in recall values of the FastText results, since this problem has a difficult and tricky process due to limited and unbalanced data. Secondly, topics found manually are compared with topics predicted with ICSIBoost. Although there is a decrease in precision value for Topic-2 compared to FastText, fairly good results are obtained for other topics. 90% precision value is achieved for Topic-1, 67% precision value is achieved for Topic-3 and 75% precision value is achieved for Topic-4. Also, we get better recall and F1-Score values with ICSIBoost for all topics compared to FastText. We observed that the dataset used in FastText's tag prediction experiments [18] is considerably larger than the dataset used in BoosTexter [19]. For this reason, we think that BoosTexter may have performed better than FastText on small data. In order to further increase the success rate, it is aimed to increase the number of SDD items as future work.

References

1. Northrop, L.M.: SEI's software product line Tenets. IEEE Softw. **19**(4), 32–40 (2002)
2. Almeida, E., et al.: Domain implementation in software product lines using OSGi. In: The 7th International Conference on Composition-Based Software Systems (ICCBSS), Madrid, Spain (2008)
3. Yu, L., Ramaswamy, S.: A configuration management model for software product line. INFOCOMP J. Comput. Sci. **5**(4), 1–8 (2006)

4. Lee, S., Choi, H.: Software component reusability measure in component grid. In: 11th International Conference on Advanced Communication Technology, pp. 576–578 (2009)
5. Rathee, A., Chhabra, J.: A multi-objective search based approach to identify reusable software components. J. Comput. Lang. **52**(3), 26–43 (2019)
6. Shatnawi, A., Shatnawi, H., Saied, M., Al Shara, Z., Sahraoui, H., Seriai, A.: Identifying software components from object-oriented APIs based on dynamic analysis. In: Proceedings of the 26th Conference on Program Comprehension (ICPC), pp. 189–199 (2018)
7. Rathee, A., Chhabra, J.K.: Mining reusable software components from object-oriented source code using discrete PSO and modeling them as Java Beans. Inf. Syst. Front. **22**(6), 1519–1537 (2020). https://doi.org/10.1007/s10796-019-09948-4
8. Padhy, N., Panigrahi, R., Satapathy, S.C.: Identifying the reusable components from component-based system: proposed metrics and model. In: Satapathy, S.C., Bhateja, V., Somanah, R., Yang, X.-S., Senkerik, R. (eds.) Information Systems Design and Intelligent Applications. AISC, vol. 863, pp. 89–99. Springer, Singapore (2019). https://doi.org/10.1007/978-981-13-3338-5_9
9. Thao, C., Munson, E.V., Nguyen, T.N.: Software configuration management for product derivation in software product families. In: 15th Annual IEEE International Conference and Workshop on the Engineering of Computer Based Systems (ECBS 2008), pp. 265–274 (2008)
10. Haris, M., Kurniawan, T., Ramdani, F.: Automated features extraction from software requirements specification (SRS) documents as the basis of software product line (SPL) engineering. JITeCS (J. Inf. Technol. Comput. Sci.) **5**(3), 279–292 (2020)
11. Hamza, M., Walker, R.: Recommending features and feature relationships from requirements documents for software product lines. In: IEEE/ACM 4th International Workshop on Realizing AI Synergies in Software Engineering, pp. 25–31 (2015)
12. Wein, S., Briggs, P.: A fully automated approach to requirement extraction from design documents. In: IEEE Aerospace Conference (2021)
13. Akin, A., Akin, M.: Zemberek, an open source NLP framework for Turkic Languages, Structure, pp. 1–5 (2007)
14. Arun, R., Suresh, V., Veni Madhavan, C.E., Narasimha Murthy, M.N.: On finding the natural number of topics with latent Dirichlet allocation: some observations. In: Zaki, M.J., Yu, J.X., Ravindran, B., Pudi, V. (eds.) PAKDD 2010. LNCS (LNAI), vol. 6118, pp. 391–402. Springer, Heidelberg (2010). https://doi.org/10.1007/978-3-642-13657-3_43
15. Ignatenko, V., Koltsov, S., Staab, S., Boukhers, Z.: Fractal approach for determining the optimal number of topics in the field of topic modeling. J. Phys. Conf. Ser. **1163**, 012025 (2019)
16. Krasnov, F., Sen, A.: The number of topics optimization: clustering approach. Mach. Learn. Knowl. Extraction **1**(1), 416–426 (2019)
17. Zhao, W., et al.: A heuristic approach to determine an appropriate number of topics in topic modeling. BMC Bioinform. **16**, 08 (2015)
18. Joulin, A., Grave, E., Bojanowski, P., Mikolov, T.: Bag of tricks for efficient text classification. In: Proceedings of the 15th conference of the European Chapter of the Association for Computational Linguistics, Short Papers, vol. 2, pp. 427–431 (2017)
19. Schapire, R., Singer, Y.: BoosTexter: a boosting-based system for text categorization. Mach. Learn. **39**, 135–168 (2000)
20. Sechidis, K., Tsoumakas, G., Vlahavas, I.: On the stratification of multi-label data. In: European Conference on Machine Learning and Principles and Practice of Knowledge Discovery in Databases, pp. 145–158 (2011)
21. Alpaydin, E.: Introduction to machine learning. (2014). https://kkpatel7.files.wordpress.com/2015/04/alppaydin_machinelearning_2010.pdf. Accessed 07 Oct 2021. (Original work published 2010)

Code Comprehension for Read-Copy-Update Synchronization Contexts in C Code

Endre Fülöp, Attila Gyén, and Norbert Pataki(✉)

Department of Programming Languages and Compilers, Faculty of Informatics, Eötvös Loránd University, Budapest, Hungary
patakino@elte.hu

Abstract. The Read-Copy-Update (RCU) mechanism is a way of synchronizing concurrent access to variables with the goal of prioritizing read performance over strict consistency guarantees. The main idea behind this mechanism is that RCU avoids the use of lock primitives while multiple threads try to read and update elements concurrently. In this case, elements are linked together through pointers in a shared data structure. RCU is used in the Linux kernel, but there are user-space libraries which implement the technique as well. One of the user-space solutions is librcu which is a C language library. We describe a code comprehension framework for easing the development of RCU solutions by aiding the programmer visually with visual cues, structurally with code-lens and can help her verify basic assumptions by tracking constraints of executions paths coming from control-flow information.

Keywords: Code comprehension · C programming language · RCU · Static analysis

1 Introduction

Read-copy-update (RCU) mechanism is used for synchronizing memory access in a way that guarantees deterministic read-access even during concurrent writes to the same memory region [4]. Unsyncronized access from multiple threads can lead to the evaluation of completely unexpected values, which in turn almost negates the programmers ability to reason about possible outcomes [7].

There are multiple families of solutions to this problem. One traditional solution is locking, where multiple threads are sequentially ordered at runtime, thus accesses to a memory region are mutually exclusive among threads. This can, however, lead to performance degradations, live- and deadlock problems. Locking solutions use syncronization primitives like mutexes and various kinds of locks [5]. Another possible solution is lock-free programming, where synchronization is solved without explicit exclusion, eliminating most locking issues [6]. Many lock-free solutions use memory barriers and atomic variables. RCU is a solution of higher abstraction level than those mentioned before. RCU can be implemented

S. Bourennane and P. Kubicek (Eds.): ICGDA 2022, LNDECT 143, pp. 187–200, 2022.
https://doi.org/10.1007/978-3-031-08017-3_17

in the kernel- or in the user-space. Linux kernel uses data structures with RCU implementation since 2002 [9].

RCU can also be implemented in the user-space, one such library is *liburcu* written in C [3]. In order to provide synchronization using the liburcu library, the user must intersperse the application code with calls to library functions. In effect the side-effects of these invocations produce a context along the execution paths, where accesses to a memory region is guaranteed to have desired properties. To help the comprehension of the synchronization provided by the library, we have devised a visualization technique. The goal of the proposed technique is to provide the users of the library a visual and interactive way of exploring the code, thus facilitating the correct and intended usage of the library. There is no silver bullet in software engineering [2], however, visualization is an important aspect.

This rest of this paper is organized as follows. In Sect. 2, we provide a brief overview on related work. We present the Userspace RCU implementation and our static analysis methods in Sect. 3. We show our visualization method in Sect. 4. Section 5 provides an explanation how the backend analysis techniques are used in our tool. We present our future work in Sect. 6, and finally, this paper concludes in Sect. 7.

2 Related Work

Visualizing concurrency aspects of programs can have the goal of assessing performance aspects of a particular solution [11]. One category of tools used to measure performance are sampling- and instrumenting profilers which are for both single- and multithreaded programs. These profilers produce aggregated performance statistics and/or traces of events which can be used for detailed performance analysis. These statistics are consequently converted into visual representations like barcharts and flamegraphs to provide an overview and highlight the proportions of each program parts contribution to a given metric. Compared to these visualizations, we propose a technique based on static analysis instead of dynamic profiling to reason about the structure of the RCU implementation. Another important aspect is that the analysis done by the RCU visualization technique is more qualitative in nature.

3 Userspace RCU Implementation

3.1 RCU Overview

RCU is implemented in program code as a set of API calls (free function calls in case of the *liburcu* C library), which implements concurrent publication of modifications on shared data, subscription for insertion into shared data structures, waiting for readers to complete their executions and finally to maintain different versions of the same data [4]. This API is geared towards read-heavy uses, where updates of values and structured data are relatively less frequent, and where consistency guarantees are not critical. Memory usage is another concern, as multiple version of the same data can lead to overuse.

Concurrent access to variables is done by associating regions of code with parts of programs executions, which read shared values (readers) [8]. These sections are called read critical sections. Read critical sections interact with synchronization points, which are usually used as part of the update part of the program executions (updaters). Readers subscribe to a specific version of the data they are reading, which is the one available at the beginning of the critical section. The end of a critical section is explicit in the code, which is needed for the updaters to detect if there is no more reading activity for a specific piece of data. Read critical sections does not enforce ordering inside a single section, nor do multiple sections between each other.

3.2 RCU Contexts in liburcu

Userspace-implemented RCU library librcu is a compile- and link-time solution for using RCU primitives in arbitrary C software without depending on kernel features of the operating system (OS) [12]. The library supports multiple implementations of the RCU semantics, the API consists of free functions with prefixes corresponding to the name of the technique (urcu) and the implementation technique (i.e. mb for memory barrier, qsrb for quiescent state-based reclamation or signal for using posix signals). By default, API calls are implemented as external linkage functions, and the generated IR code therefore contains explicit references to the mentioned free functions. Using optimizations which cause the functions to be inlined will render the solution described here unusable. Inlining small functions can be the result of link-time optimization or by defining the URCU_INLINE_SMALL_FUNCTIONS preprocessor symbol before including the library headers.

Another limitation is that debug information must be generated alongside with the IR code. An example of an API function which is used for opening a read-side critical section by using memory barriers as implementation is urcu_mb_read_lock().

There are two API functionalities, which must be used in pairs. For registering threads, one would used the urcu_<flavor>_register_thread() and urcu_<flavor>_unregister_thread(). These are used in a non-nested way (calling register while already registered is an error), but the other pair of API functions signifying the read-side critical sections can be nested indefinitely. These are the urcu_<flavor>_read_lock() and urcu_<flavor>_read_unlock() functions. The solution presented here is tailored towards the nestable usage, and can be extended to consider the non-nestable case. There are API calls which can only be safely used inside the context of registered threads (the majority of the librcu API) and there are API calls, which have special meaning when inside a read-side critical section (like defer_rcu() or synchronize_rcu()). The intended usage of the solution presented here is to provide information about the potential execution paths that are potentially enclosed in the mentioned API calls. It would help the software's discoverability, changeability, and maintainability to know which part of the code potentially contributes to the synchronization structure.

3.3 Context Detection in LLVM IR Code

LLVM Project is a collection of compiler and toolchain technologies, which aims to provide a modular approach to translation, optimization and linking of programming languages [1]. One of the main contribution of the project is a common intermediate representation between high-level programming languages and binary, called LLVM Intermediate Language (LLVM IR), along with tools to translate high level programming languages to IR, optimize IR code, generate binary code from IR code and package binaries on multiple platforms. There are multiple frontends and backends for the IR representations. For static analysis purposes, we only consider the frontend part, as the solution represented by the authors use the IR representation with some embedded debug information to identify locations in the original source code.

Defining analyses on LLVM IR has the advantage of possible use from different host languages which can be translated to LLVM IR. The other way around is also possible, if the binary representation of a program can be transformed (referred to as lifting) into LLVM IR form, and analysis can be performed on the lifted code. Also various optimization passes can simplify and transform the original code, but link-time dependency on the API functions preserve the necessary information to still yield usable results using this technique. The following code snippet is an example of IR representation of a source with a simple loop:

```
@x = external dso_local global i32, align 4

; Function Attrs: noinline nounwind optnone uwtable
define dso_local i32 @main() #0 !dbg !9 {
entry:
  %retval = alloca i32, align 4
  store i32 0, i32* %retval, align 4
  call void (...) @urcu_begin(), !dbg !13
  br label %while.cond, !dbg !14

while.cond:               ; preds = %while.body, %entry
  %0 = load i32, i32* @x, align 4, !dbg !15
  %tobool = icmp ne i32 %0, 0, !dbg !14
  br i1 %tobool, label %while.body, label %while.end,
        !dbg !14

while.body:               ; preds = %while.cond
  call void (...) @urcu_begin(), !dbg !16
  call void (...) @f(), !dbg !18
  call void (...) @urcu_end(), !dbg !19
  br label %while.cond, !dbg !14, !llvm.loop !20

while.end:                ; preds = %while.cond
  call void (...) @urcu_end(), !dbg !23
  ret i32 0, !dbg !24
```

```
}

declare dso_local void @urcu_begin(...) #1
declare dso_local void @f(...) #1
declare dso_local void @urcu_end(...) #1
```

3.4 Context Detection Algorithm

Detecting the possible read side critical sections in user code in this solution uses static analysis. The IR code is analyzed in a flow-sensitive manner that leads to an over-approximation of the possible RCU contexts. The over-approximation is deemed useful by the authors, as it leads to bounded execution time of the analysis depending only control structure of the analyzed code, and is independent of the type representation set sizes used in the code.

Data-flow analysis is a technique used in compiler-technology to extract information about not only the static structure of the program that is defined by the grammar and is captured by the AST representation, but also information about the dynamic behavior of the program. Flow sensitivity means that possible execution paths allowed by the control-flow structure of the program are analyzed. Constraints along paths are not taken into consideration, therefore flow-sensitive over-approaches over-estimate the set of actually possible execution paths of a program. Path-sensitive approaches approximate the dynamic behavior of the program better generally, but have higher cost. Quick analysis time is essential to provide low-latency results for editors of the code, thus we have chosen a flow-sensitive approach.

The iterative algorithm of forward dataflow analyses uses reverse postorder traversal of the control-flow graph (CFG) elements in case of forward analysis in order for performance reasons. The solution presented here (see Fig. 1) uses a modified iterative forward dataflow analysis. The reverse postorder visitation of the CFG elements is no longer a performance factor, but is an important part of the correctness. The modified forward dataflow technique is more similar to a may analysis, therefore one of the closest widespread dataflow usecase would be the reaching definitions problem [12].

The dataflow facts are associated to multiple levels of the CFG. There are function level facts, basic block level facts and instruction level facts. The analysis presented is interprocedural in nature and employs rudimentary context-sensitivity, taking the callstack of the analyzed functions into account. This information is currently not emitted to the visualization part. The basic dataflow facts are used to associate a list of interesting source-locations to instructions in the IR code. The context starter and ending functions are given as parameters to the algorithm and identified by name inside the IR code. The LLVM IR `call` instruction encodes the API calls.

A source location is interesting if it belongs to an instruction that is either a context starting or a context ending function call. There are four other interesting locations. When entering a function, the entry point to an interprocurally

ALGORITHM 1: Flow sensitive context detection

```
CallContextWorklist functions_to_explore
Set<Function> active_functions
functions_to_explore.addInitialFunction(<main function>)
AnalysisResult result
while !functions_to_explore.empty(), do
  Function f = functions_to_explore.takeNextFunction()
  active_functions.add(f)
  FunctionDataflowFacts old_function_facts = result.getFunctionFacts(f)
  old_function_facts.initialize_function_facts()
  old_function_facts.initialize_basic_block_entry_facts()

  BasicBlockList block_worklist = reverse_postorder_visit(f)
  while !block_worklist.empty(), do
    BasicBlock b = block_worklist.takeNextBasicBlock()
    BBDataflowFacts old_facts = result.getBasicBlockExitFacts(b) // copy of the old facts for the fixed-point iteration
    old_facts.update_with_predecessor_facts(result.getBasicBlockEntryFacts(b))
    old_facts.update_with_interprocedural_facts()

    for each i in forward_iterate_instructions(b), do
      if should_interprocedurally_analyze(i), do
        result.getFunctionFacts(i).initilizeWithCallContext(old_function_facts)
        b.getBBGlobalState().markFuntionEnter()
        // apply this algorithm to the called function
        result.detectContextInCalledFunctions(f, i, result, old_function_facts)
        b.getBBGlobalState().markFuntionLeave()
      end
      result = apply_transfer_function(i, result)
      if old_facts != result.getBasicBlockExitFacts(b), do
        block_worklist.add(b.successors())
      end
    end

    if old_function_facts != result.getFunctionFacts(f), do //changed the function level facts
      functions_to_explore.add(f.callers())
    end

    active_functions.remove(f)
  end
```

Fig. 1. Flow sensitive context detection

analyzed call on the caller side, the first instruction with a source location inside the callee. When returning from a function the last instruction with a source location on the callee's side (note that this location is the over-approximation of all possible exit-points of the function), and exit point on the callers side.

The transfer function (see Fig. 2) saves the interesting locations (the instructions that can be used to get the locations), by appending them to the basic block level global fact, but only if this global fact is does not already contain them. In addition, if a context ending API call is detected, the exit state of the instruction set to the current global state of the basic block. The reverse postorder visitation guarantees, if a context starting instruction then happens to precede a context ending one, there is path in the CFG from the starter to the ending one. The set-like nature of the list in turn allows for the halting of the fixed-point algorithm in finite steps, as there are a finite amount of interesting locations inside a program.

ALGORITHM 2: Transfer function for context detection

```
Instruction i
BBDataflowFacts block_of_i
BBGlobalFacts global_facts = block_of_i.getBBGlobalState()

if (i.isContextStartingFunction() or i.isContextEndingFunction()), then
   if (!global_facts.contains(i)), then
      global_facts.append(i)
   end
end

if (I.isContextEndingFunction()), then
   block_of_i.setFactsForInstruction(i, global_facts)
end
```

Fig. 2. Transfer function for context detection

The meet function (see Fig. 3) is responsible for merging the exit states of multiple incoming dataflow facts. This is defined as the concatenation of the dataflow fact lists in a manner, that guarantees uniqueness of elements inside the resulting list, and the preservation of relative ordering among the interesting locations. In order to facilitate the transfer and meet functions the dataflow fact lists are implemented with a `SetVector` data structure, which allows for uniqueness-preserving insertion, and insertion-ordered iteration at the same time (implemented via a vector and a set).

After generating the dataflow facts for every instruction, the reporting of the contexts is done by finding the instructions which encode context ending API calls, and matching it with every context beginning API call locations inside its dataflow facts list. Then emitting the ranges from the beginning to the end API calls with the interspersed locations produced by the function enter and leave events of the interprocedural part of the analysis produces source-location ranges. Currently only the line information is used by the visualization part. The debug information emitted by the LLVM IR generator makes it hard to reason about the end of the source-range of an instruction.

ALGORITHM 3: Meet function for context detection

```
InstructionDataflowFacts fact_list_first
InstructionDataflowFacts fact_list_second

InstructionDataflowFacts result_fact_list = fact_list_first
for each instruction in fact_list_second, do
   if (!result_fact_list.contains(instruction)), do
      results_fact_list.append(instruction)
   end
end

// InstructionDataflowFacts is a datatype which preserves insertion order and uniqueness of elements
// It is implemented with SetVector data structure
```

Fig. 3. Meet function for context detection

The following examples present different constructs. We show what is the original code snippet and what is JSON formatted result. We present an if-statement branched code and its result, a code snippet includes a loop and its result. Finally, an interprocedural approach is presented. The JSON formatted results encode the list of contexts in the form of begin- and end-keyed source location pairs. The total number of contexts found is the multiplicity of the outermost JSON array.

```
1  void urcu_begin();
2  void urcu_end();
3  void f();
4  extern int x;
5
6  int main() {
7     urcu_begin();
8     if ( x ) {
9        f();
10       urcu_end();
11    } else {
12       urcu_end();
13       f();
14    }
15 }
```


```
[[{"begin":"branch.c@8:3", "end":"branch.c@11:5"}],
 [{"begin":"branch.c@8:3", "end":"branch.c@13:5"}]]
```


```
1  void urcu_begin();
2  void urcu_end();
3  void f();
```

```
 4  extern int x;
 5
 6  int main() {
 7    urcu_begin();
 8    while ( x ) {
 9      urcu_begin();
10      f();
11      urcu_end();
12    }
13    urcu_end();
14  }
```


```
[[{"begin":"loop.c@8:3", "end":"loop.c@12:5"}],
 [{"begin":"loop.c@10:5", "end":"loop.c@12:5"}],
 [{"begin":"loop.c@8:3", "end":"loop.c@14:3"}],
 [{"begin":"loop.c@10:5", "end":"loop.c@14:3"}]]
```


```
 1  void urcu_begin();
 2  void urcu_end();
 3  extern int x;
 4
 5  void f() {
 6    if ( x )
 7      urcu_end();
 8  }
 9
10  int main() {
11    urcu_begin();
12    f();
13    urcu_end();
14    return 0;
15  }
```


```
[[{"begin":"interproc.c@12:3", "end":"interproc.c@14:3"}],
 [{"begin":"interproc.c@12:3", "end":"interproc.c@13:3"},
  {"begin":"interproc.c@6:7", "end":"interproc.c@7:5"}]]
```


4 Visualizing Userspace RCU Contexts

4.1 Visualization of the Result

Our goal was not only to extract the output from the raw C/C++ code with the appropriate data, which holds where the user started an RCU range and where it ends, but also wanted to visualize it in a way in which the programmers or the users can have better understanding what happened in the background during the runtime. Several options have been examined to find the best editor or the best tool for this job.

First of all, we aimed to work with Visual Studio Code because it is one of the most popular code editor nowadays. Therefore, many developers could have used our visualization tool. Unfortunately, Visual Studio Code does not support to edit the style of the editor in a way we needed.

Second approach was the Monaco online editor which fully met the needs – or at least the majority – we were looking for.

4.2 Monaco Editor

Monaco Editor is maintained by Microsoft and available worldwide for free [10]. It has a playground with full of interactive examples and provides wide access to the editor and it supports feature like colorize the editor line-by-line, add different error and warning messages or add a hover message when the cursor is hovered over the text. Doing all this with JavaScript programming language for the dynamic parts, CSS for styling and HTML to build the raw frame. It gives full access to the Document Object Model (DOM) supplemented by its own special elements. However, it sets up some limitations.

The editor separates the languages into different files whose content must be predefined. This means there is no option to generate the HTML on-the-fly in JavaScript and then append the new items to existing ones. Same is true for the CSS. Only the predefined classes can be used. The size or the style of the fonts cannot be changed even inside the CSS. All this means the user needs to know in advance which classes and tags he/she wants to use. This is not the most convenient method because, for long code, it may not be predictable, for example, how many different sections need to be displayed in different colors and every time the user wants a change he/she has to add new elements manually and run the code again.

When working with code ranges, the appropriate list of ranges has to be implemented in the editor. This was the other problem. Monaco editor cannot parse the output of the analyzed code, despite it is a JSON formatted, however, its content is inappropriate. There is no predefined API with which to solve this issue.

In order to make it easier to solve these problems, we created a tool, which is covered in more detail in the following section.

5 Backend Analysis Techniques

As it was mentioned, we started to write a range-parser tool which can process the extracted data from the JSON file and can display it in a format that matches the Monaco editor needs. It is important to mansion that this tool does not provide a solution to the problem with dynamic content for now.

Consider the following code snippet:

```
1  void urcu_begin();
2  void urcu_end();
3  void f();
4
5  int main() {
6    urcu_begin();
7    f();
8    urcu_end();
9    urcu_end();
10   return 0;
11 }
```

From the above code the RCU code parser will extract the following output:

```
[[{"begin": "urcu3.c@7:3", "end": "urcu3.c@9:3"}],
 [{"begin": "urcu3.c@7:3", "end": "urcu3.c@10:3"}]]
```

As shown the output is a list, containing lists of objects where every object holds key-value pairs, namely the begin which indicates the start and end which indicates the end of the RCU block.

The range-parser is written with the help of Python programming language and its output is a Python dictionary object which is equivalent to JavaScript's associative-array and for this reason it can be easily assigned to a variable inside the online editor.

The JSON filename which holds the extracted ranges should be passed as an input parameter when executing the range-parser. It breaks down the list into stand-alone objects and assembles a new object which holds range names as keys and a list to every range. The lists have four values: start row's row and column number and the end row's row and column number. The reduced list from the range-parser is the following:

```
{range79: [7, 0, 9, 0], range710: [7, 0, 10, 0]}
```

The editor cares about only the row numbers and will color the full row. For that reason, we added a constant 0 value for every column - column in this case represents the number of characters in the given row. The name of the ranges always starts with a range prefix followed by the concatenated start and end row number.

This object now can be easily assigned to a variable inside the online editor:

```
const range = {range79: [7, 0, 9, 0],
               range710: [7, 0, 10, 0]}
```

Monaco editor defines an `editor` object with help of which the user can interact with the editor itself with the `deltaDecorations()` function. It has two input parameters, the first one is an object that contains the old decorations - these will be deleted. The second one holds the new decorations which will be applied. Both of them is a list of object where are predefined key names like range for the new range and options. In the latter, we can refer to a CSS class as string or add hover messages. First need to setup the editor and the used programming language:

```
var editor =
  monaco.editor.create(
    document.getElementById('container'), {
      value: cppCode, language: 'cpp', glyphMargin: true
  });
```

The cppCode holds the raw C++ code itself copied from editor which was seen in the previous section. Now we are able to add decorations:

```
var decoratorList = [{
  range: new monaco.Range(...range.range79),
  options: {
    isWholeLine: true,
    className: 'range79',
    hoverMessage: [{value: 'Belongs to range 7-9'}]
}},
{
  range: new monaco.Range(...range.range710),
  options: {
    isWholeLine: true,
    className: 'range710',
    hoverMessage: [{value: 'Belongs to range 7-10'}]
  }
}];

var decorations = [];
decorations = editor.deltaDecorations(decorations,
                                       decoratorList);
```

The real time visualization makes the editor available to parse a stream and format the rows in the editor.

To make it even clearer for the user where exactly the RCU blocks are taking place we added a range selector field for every RCU range in the code with some several options:

- All range
- Range: [from]-[to]
- No range

For longer code snippets whose start and end points do not fit on the display, we have assigned a hover message that shows how long the particular part lasts from.

6 Future Work

The detection algorithm can be extended to provide the context-sensitive information to the visualization part in a form that enhances comprehension. The multiple layers of dataflow fact information give space for querying functionality. Targeted requests could be made by the visualization part to the analysis engine to provide for example results with only a given call-context. This querying mechanism could enhance the performance of the analysis as well, as targeted inlining would be possible during analysis. Conditional statements could be taken into consideration when calculating interesting points. Other analysis techniques could then be used in conjunction to prune the possible set of contexts reported by the current pure flow-sensitive solution. We want to make the range-parser more sophisticated and available to generate the full source code which can be easily copy-pasted into Monaco online editor. Thus avoiding the difficulty that occurs when there has been some change in the code the ranges have changed or new ranges were added that needs to be highlighted in a different way. Offer several options for the user from he/she can select the ones he/she want to see in the editor, like hover messages or the range selection dropdown list.

7 Conclusion

Despite RCU is a very powerful mechanism and in a sense simplifies thread handling in order for someone to understand what is going on in the background, a deeper understanding of the topic is required. The visualization tool does not answer all questions, but it helps to understand the background processes better. The flow-sensitive analysis approach presented here is a scalable method for gaining an overview about the synchronization aspects of the software. The modular nature of the approach lends itself to distributed use.

References

1. Babati, B., Horváth, G., Májer, V., Pataki, N.: Static analysis toolset with Clang. In: Proceedings of the 10th International Conference on Applied Informatics, pp. 23–29 (2017)
2. Danisovszky, M., Nagy, T., Répás, K., Kusper, G.: Western canon of software engineering: the abstract principles. In: 2019 10th IEEE International Conference on Cognitive Infocommunications (CogInfoCom), pp. 153–156 (2019). https://doi.org/10.1109/CogInfoCom47531.2019.9089999
3. Desnoyers, M., McKenney, P.E.: Userspace RCU. https://liburcu.org/
4. Desnoyers, M., McKenney, P.E., Stern, A.S., Dagenais, M.R., Walpole, J.: User-level implementations of Read-Copy update. IEEE Trans. Parallel Distrib. Syst. **23**(2), 375–382 (2012). https://doi.org/10.1109/TPDS.2011.159
5. Drocco, M., Castellana, V.G., Minutoli, M.: Practical distributed programming in C++. In: Proceedings of the 29th International Symposium on High-Performance Parallel and Distributed Computing, pp. 35–39, HPDC 2020. Association for Computing Machinery, New York, NY, USA (2020). https://doi.org/10.1145/3369583.3392680

6. Hart, T.E., McKenney, P.E., Brown, A.D., Walpole, J.: Performance of memory reclamation for lockless synchronization. J. Parallel Distrib. Comput. **67**(12), 1270–1285 (2007). https://doi.org/10.1016/j.jpdc.2007.04.010, https://www.sciencedirect.com/science/article/pii/S074373150700069X, Best Paper Awards: 20th International Parallel and Distributed Processing Symposium (IPDPS 2006)
7. Márton, G., Szekeres, I., Porkoláb, Z.: Towards a high-level C++ abstraction to utilize the Read-Copy-Update pattern. Acta Electrotechnica et Informatica **18**(3), 18–26 (2018). https://doi.org/10.15546/aeei-2018-0021
8. McKenney, P.E.: Is parallel programming hard, and if so, what can you do about it? (release v2021.12.22a) (2021). https://arxiv.org/abs/1701.00854
9. McKenney, P.E., Walpole, J.: What is RCU, fundamentally? (2007). https://lwn.net/Articles/262464/
10. Microsoft: Monaco editor. https://microsoft.github.io/monaco-editor/
11. Porkoláb, Z., Brunner, T.: The codecompass comprehension framework. In: Proceedings of the 26th Conference on Program Comprehension, pp. 393–396, ICPC 2018. Association for Computing Machinery, New York, NY, USA (2018). https://doi.org/10.1145/3196321.3196352
12. Umann, K., Porkoláb, Z.: Detecting uninitialized variables in C++ with the Clang Static Analyzer. Acta Cybernetica, November 2020. https://doi.org/10.14232/actacyb.282900, https://cyber.bibl.u-szeged.hu/index.php/actcybern/article/view/4100

Author Index

S. Bourennane and P. Kubicek (Eds.): ICGDA 2022, LNDECT 143, pp. 201–202, 2022.
https://doi.org/10.1007/978-3-031-08017-3

Printed in the United States
by Baker & Taylor Publisher Services